Frequency Selective Rasorber Designs for Stealth Applications

This book offers a comprehensive guide to the principles, design methodologies, and practical implementations of frequency selective rasorbers (FSRs) with a focus on radar cross-section (RCS) reduction for modern stealth technologies. It clears the fundamentals of FSRs and absorptive structures while bridging the gap between theory and real-world application. Through a structured and accessible approach, readers will gain a deep understanding of how to design, simulate, fabricate, and evaluate narrowband and broadband FSRs. The book also explores cutting-edge low-RCS antenna designs co-integrated with FSRs, serving as both a foundational text and a motivation for new research directions.

This book:

- Contains in-depth discussion of the theoretical foundations and classifications of frequency selective rasorbers.
- Offers a systematic organization of various FSR types across dedicated chapters for clarity and ease of understanding.
- Presents step-by-step design procedures for both narrowband and broadband rasorber structures.
- Explains practical insights into RCS reduction techniques, including co-designed low-RCS antenna systems.
- Provides detailed coverage of fabrication and measurement processes for validating rasorber and stealth system performance.

It is primarily written for senior undergraduates, graduate students, and academic researchers in radio frequency (RF), microwave, and antenna engineering.

Frequency Selective Rasorber Designs for Stealth Applications

Mehran Manzoor Zargar, Archana Rajput,
and Kushmanda Saurav

CRC Press
Taylor & Francis Group
Boca Raton London New York

CRC Press is an imprint of the
Taylor & Francis Group, an **informa** business

Designed cover image: Mehran Manzoor Zargar, Archana Rajput, Kushmanda Saurav

First edition published 2026
by CRC Press
2385 NW Executive Center Drive, Suite 320, Boca Raton FL 33431

and by CRC Press
4 Park Square, Milton Park, Abingdon, Oxon, OX14 4RN

CRC Press is an imprint of Taylor & Francis Group, LLC

ISBN: 978-1-041-08216-3 (hbk)
ISBN: 978-1-041-08217-0 (pbk)
ISBN: 978-1-003-64432-3 (ebk)

DOI: 10.1201/9781003644323

Typeset in Sabon
by Apex Covantage, LLC

Contents

Preface

In the modern global world, the progress of communication technologies is unending. The transmission of electromagnetic waves forms the foundation of wireless communication systems. With the advancing technology, there is a significant interest among researchers in selectively controlling the propagation of electromagnetic waves. Of particular importance is the absorption of these waves, especially in stealth applications where enhancing target secrecy by minimizing its radar cross-section (RCS) is crucial. Additionally, electromagnetic absorption plays a vital role in applications focused on shielding and reducing electromagnetic interference (EMI).

The periodic structures of frequency selective absorbers (FSAs) have been developed over the years, which causes the absorption of incident electromagnetic waves. A unique feature of FSS-based absorbers lies in its thin, lightweight, and frequency-selective properties over the conventional absorbers, which makes them easy to use and handle in the stealth technology. To address the limitation of communication blockade in FSA, another structure known as a frequency selective rasorber (FSR) or simply rasorber has been developed. The frequency selective rasorber (FSR) is a periodic microwave structure that is capable of absorbing the incident electromagnetic (EM) wave in a particular frequency band while providing a passband to the incident wave at some other frequency spectrum. FSR is an advanced version of a frequency selective absorber and provides an additional transmission band along with the absorption band. Being thin, light, and frequency-engineered, these absorber structures are more useful than the conventional bulky absorbers in various applications like shielding and interference/RCS reduction.

In this book, we have attempted to present comprehensive and systematic guidance towards designing, fabrication, and application of various categories of frequency selective rasorbers. The book starts with the concept of single-layer narrowband rasorber designs and sequentially introduces the multilayer broadband rasorber designs followed by the RCS reduction application of the FSR. This book can serve as a handbook for readers in understanding the concept and designing of an FSR together with the RCS reduction applications and will further motivate them to develop new ideas.

The book has been divided into eight chapters. A brief overview of the frequency selective rasorber is discussed in Chapter 1. Starting from EM absorption, the chapters continue to discuss frequency selective surface based absorbers/rasorbers and the state-of-art designs. Chapter 2 presents a design of single-layer narrow band rasorber with single absorption band and dual transmission band. The designing steps are laid down along with equivalent circuit modeling and experimental measurements. Chapter 3 presents the design of layer-flexible rasorber with dual and triple absorption bands along with two transmission windows. The experimentation for the multiband rasorbers involves the test for conformal features. In Chapter 4, an A-T (absorption-transmission) FSR is discussed where a transmission band at a higher frequency is achieved besides the lower broadband absorption. The FSR is designed by first studying and then combining the individual designs of the broadband absorber and bandpass frequency selective surface (FSS). In Chapter 5, a compact T-A-T (Transmission-Absorption-Transmission) type of FSR is presented with dual transmission bands located at the two sides of a broad absorption band. In Chapter 6, a design of a compact frequency selective rasorber exhibiting absorption-transmission-absorption characteristics and having high passband selectivity has been illustrated. Further, in Chapter 6, a low radar cross-section antenna is realized by integrating a patch antenna with the proposed FSR, which achieves an average out-of-band monostatic/bistatic radar cross-section (RCS) reduction, while maintaining the other antenna parameters. Next in Chapter 7, a novel design strategy for realizing a miniaturized A-T-A-T (absorption-transmission-absorption-transmission) type of frequency selective rasorber (FSR) having dual-band transmission capabilities is presented. Chapter 8 discusses a novel strategy for realizing a reduction in radar cross-section (RCS) of a dual-polarized crossed dipole antenna. A compact and polarization-insensitive band notched absorber, also known as absorptive frequency selective reflection (AFSR) structure, is proposed and then integrated with a dual-polarized crossed dipole antenna for achieving monostatic as well as bistatic RCS reduction as compared to the conventional reflector backed counterpart.

Mehran Manzoor Zargar
Archana Rajput
Kushmanda Saurav

Biographies

Mehran Manzoor Zargar received the B.Tech. degree in Electronics and Communication Engineering from the University of Kashmir, India, in 2014; the M.Tech. degree in Microelectronics from the National Institute of Technology Srinagar, India, in 2018; and the Ph.D. degree from the Department of Electrical Engineering, Indian Institute of Technology Jammu in 2023, where he is currently working as a Postdoctoral Research Associate, under the Core Research Grant Project funded by the Anusandhan National Research Foundation (ANRF), Government of India. He was an Early Postdoctoral Fellow with the Department of Electrical Engineering, IIT Jammu from February 2023 to March 2023. He has published several research papers in esteemed journals and conferences. He has served as reviewers for many reputed journals like IEEE TAP, IEEE AWPL, IEEE Access Wiley MOTL. His research interest includes FSS-based rasorbers and low RCS antenna systems. Dr. Zargar is IEEE member and has served as Chair for IEEE AP-S Student Branch Chapter of IIT Jammu for 2022 and 2023.

Archana Rajput received the B.Tech. Degree in Electronics Engineering from Kamla Nehru Institute of Technology, Sultanpur, affiliated under Uttar Pradesh Technical University Lucknow, in 2009; the M. Tech degree in microwave electronics from Delhi University, South Campus, in 2012; and the Ph.D. degree from the Department of Electrical Engineering from Indian Institute of Technology, Kanpur, India in 2018. She is currently Assistant Professor with the Department of Electrical Engineering, Indian Institute of Technology Jammu, India. She has published several research papers in esteemed journals and conferences. One of her journal papers, entitled "Design of a Two-Dimensional Metamaterial Cloak with Minimum Scattering Using a Quadratic Transformation Function," has been featured on the Cover Page of *AIP Journal of Applied Physics*. Her current research interests include transformation electromagnetics, cloaking/invisibility, metamaterials, antennas, and frequency selective surfaces.

Kushmanda Saurav received the B.E. degree in electronics and telecommunication from Bengal Engineering and Science University, Shibpur,

India, in 2009; the M.Tech. degree in Electronics and Communication Engineering from IIT Roorkee, India, in 2011; and the Ph.D. degree in Electrical Engineering from IIT Kanpur, India, in 2017. He worked as a Postdoctoral Fellow with the Royal Military College Canada. He is currently Associate Professor with the Department of Electrical Engineering, IIT Jammu, India. His research interests include microwave and millimeter wave antennas and frequency-selective surfaces. He has published more than 70 research papers in esteemed journals and conferences. He has served as reviewer in IEEE, IET, and Wiley Journals. He is actively involved in the professional activities of the IEEE. He is currently the faculty advisor of IEEE MTTS and APS Student Branch Chapters of IIT Jammu. He has the experience of supervising M.Tech. and Ph.D. students and undertaking sponsored projects.

Chapter 1

Introduction

Electromagnetic (EM) absorbers are engineered materials or structures that effectively attenuate incident EM waves by suppressing both reflection and transmission. Classic absorber configurations, including the Salisbury screen, Jaumann absorbers, and carbon-foam-based pyramidal structures, have demonstrated high absorption efficiencies across various frequency bands. EM absorption is critically important in applications such as radar cross-section (RCS) reduction, electromagnetic interference (EMI) shielding, and secure communication systems. However, the inherent bulkiness and weight of conventional absorbers present challenges for integration into compact and lightweight systems. Significant advancements in frequency selective surfaces (FSS) have enabled the development of thin, lightweight, and highly customizable absorbers with tunable electromagnetic responses [1]. FSS-based absorbers exhibit several advantages, including structural compactness, broadband or multiband performance, and design flexibility, rendering them highly attractive for applications in stealth technology, EMI shielding, and beyond [2]. Over the past decade, research has increasingly focused on the systematic analysis, characterization, and optimization of FSS-based absorbers to meet the evolving demands of modern electromagnetic applications.

Despite their advantages, FSS-based absorbers face a critical limitation: they inherently block all incident signals, making them unsuitable for applications requiring selective electromagnetic wave transmission. This challenge has been addressed by integrating absorptive layers with frequency selective bandpass structures, resulting in a hybrid device known as the frequency selective rasorber (FSR), or simply "rasorber." Since the introduction of the FSR concept, it has attracted significant attention, and numerous design methodologies have been proposed in recent literature. This book focuses on the systematic design of FSRs tailored to different operational requirements, emphasizing enhancements in performance compared to previously reported structures. In addition, the book also presents chapters exploring the integration of FSRs with antenna systems to demonstrate their practical applicability in stealth technology.

The present chapter provides a structured overview of the FSS-based absorber/rasorber and the designs presented in this book. It begins with a

DOI: 10.1201/9781003644323-1

brief introduction to stealth fundamentals and FSS, followed by the theoretical background and literature review related to FSS-based absorbers. Subsequently, the concept of FSR is discussed, along with an in-depth survey of recent developments in the field and the associated practical challenges. The chapter then outlines the motivation behind the present book, description of the overall organization, and content structure of the book followed by summary of the key outcomes.

1.1 STEALTH TECHNOLOGY FUNDAMENTALS

Stealth technology, also known as low observable technology, encompasses a set of design principles and engineering techniques aimed at minimizing the detection, interception, and tracking of objects by various detection systems such as radar, infrared, sonar, and visual observation. Originally pioneered for military applications, stealth has evolved into a multidisciplinary domain with applications in defense, aerospace, surveillance, and electromagnetic compatibility (EMC).

The core objective of stealth technology is to reduce the RCS, which quantifies the measure of how detectable an object is by radar. A lower RCS translates to a lower probability of detection by hostile radar systems, which enhances survivability in combat environments. RCS reduction is achieved through a combination of platform shaping, material usage, and electromagnetic surface treatments.

Several fundamental approaches are enlisted for achieving stealth:

- *Geometric Shaping*: This involves designing surfaces and structures that deflect incoming radar waves away from the radar source. Examples include angled panels, serrated edges, and curved geometries that minimize specular reflections.
- *Radar-Absorbing Materials (RAMs)*: These materials absorb incident electromagnetic energy and convert it into heat, thereby preventing reflection back to the radar receiver. RAMs are frequency-dependent and require precise material selection and thickness control.
- *Active Cancellation*: This advanced technique involves generating counter-radar signals that destructively interfere with the incoming radar wave, effectively canceling the return signal.
- *FSS-Based Absorbers*: FSSs are periodic structures designed to filter or absorb electromagnetic waves in specific frequency bands. When engineered appropriately, they can contribute significantly to stealth by selectively absorbing or suppressing unwanted signals.

In recent years, integrated stealth systems have gained attention, where stealth features are embedded directly into the functional components of a platform such as antennas, radomes, or solar panels. This integration avoids the need for bulky external treatments and enhances multifunctionality.

Modern stealth designs must balance low observability with operational performance. For instance, aircraft and missile platforms must maintain aerodynamic efficiency, communication ability, and thermal management alongside reduced RCS. This trade-off necessitates adaptive or frequency selective solutions that are capable of distinguishing between hostile radar signals and friendly communication channels. This is where technologies such as frequency selective rasorbers (FSRs) come into play. FSRs provide a dual capability of absorbing unwanted radar signals while allowing critical communication or sensing signals to pass through, making them vital in next-generation stealth systems.

1.2 FREQUENCY SELECTIVE SURFACE

A frequency selective surface (FSS) is a periodic structure comprising an array of metallic patterns printed on a dielectric substrate. When illuminated by an incident electromagnetic (EM) wave, the FSS exhibits frequency selective behavior that is allowing either reflection or transmission, depending on parameters such as the geometry, dimensions of the unit cell, polarization and angle of the incident wave [3]. Due to this tunable spectral response, FSS structures have found widespread applications in diverse areas, including radome design [4–6], dichroic sub-reflectors [7], radio frequency identification (RFID) systems [8], electromagnetic lenses [9, 10], absorbers [11], cloaking devices [12, 13], sensors [14–16], polarizers [17–23], reflectarray antennas [24], and electromagnetic shielding [25, 26]. In recent years, the versatility of FSS has been extended to several emerging applications [27–42]. The electromagnetic behavior of an FSS is primarily governed by the design of its basic repeating unit, or unit cell. While unit cell shapes may vary, the underlying operational principle is typically based on resonance. When an EM wave impinges on the FSS, resonant currents are induced in the metallic patterns, with the resonant frequency determined by the unit cell's geometry, material properties, and the characteristics of the incident wave [43].

The fundamental characteristics of FSS designs are illustrated using two representative structures in Figures 1.1(a) and 1.1(b), consisting of arrays of metallic patches and complementary slots, respectively. As observed in the transmission and reflection responses in Figure 1.1(c), the patch-type FSS exhibits a bandstop behavior characterized by a pronounced dip in the transmission coefficient (S_{21}) and a corresponding peak in the reflection coefficient (S_{11}) at the resonance frequency. This behavior arises from the inductive and capacitive effects induced by the incident EM wave, which can be modeled as a series LC circuit. Conversely, the slot-type FSS shown in Figure 1.1(b) exhibits a bandpass response. As depicted in Figure 1.1(d), at the resonant frequency, the reflection coefficient (S_{11}) reaches a minimum while the transmission coefficient (S_{21}) attains a peak. This response is analogous to that of a parallel LC circuit. The comparison highlights how the geometry of the unit cell (whether patch-based or slot-based) determines whether the FSS exhibits a stopband or passband characteristic.

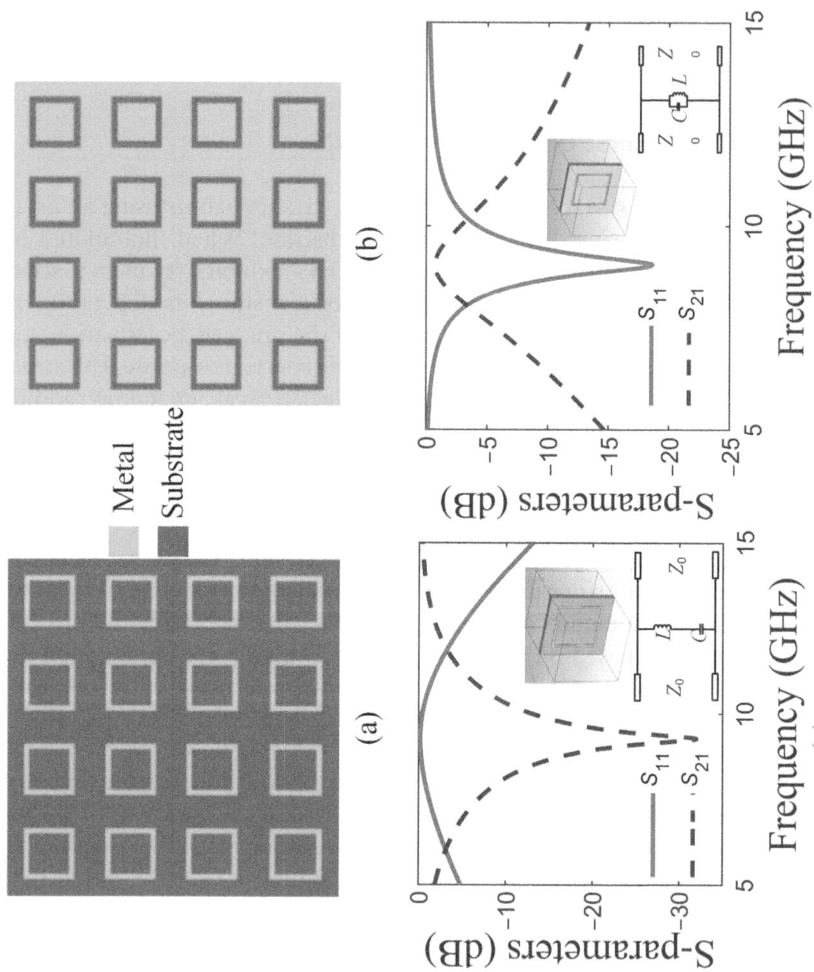

Figure 1.1 FSS structures: (a) array of patches and (b) array of slots. S-parameter responses: (c) array of patches and (d) array of slots.

1.3 FSS-BASED ABSORBERS

FSS have emerged as a pivotal technology in the design of modern electromagnetic absorbers, offering unique capabilities that address the limitations of traditional bulk absorbers. The FSS-based absorbers provide compact, lightweight, and frequency selective features, making it suitable for advanced electromagnetic (EM) applications, particularly in stealth, communication, and electromagnetic compatibility (EMC) systems.

FSS-based absorbers consist of periodic two-dimensional arrays of conductive or resistive elements printed or patterned on dielectric substrates. These periodic elements can be tailored in terms of shape, size, orientation, and arrangement to interact with specific frequency bands of the incident EM wave. The geometry and electrical properties of the FSS determine its resonant behavior, enabling designers to control the absorption characteristics precisely.

One of the most significant advantages of FSS absorbers is their spectral selectivity. While traditional absorbers indiscriminately absorb a wide range of frequencies, FSS-based absorbers can be engineered to target and attenuate specific frequency bands. This makes them particularly suitable for applications requiring narrowband or multiband absorption, such as RCS reduction at specific frequencies or eliminating interference in sensitive communication bands.

Moreover, FSS absorbers can be integrated with resistive films or lossy materials to enhance absorption performance. These hybrid structures combine the frequency selective response of the FSS with the dissipative nature of the lossy medium, thereby achieving effective absorption with minimal thickness. As a result, FSS-based absorbers are especially advantageous for conformal and embedded stealth surfaces, where size, weight, and aerodynamic profile are critical.

Over the past decade, extensive research has focused on optimizing the design methodologies, material choices, and fabrication techniques of FSS-based absorbers. Numerous topologies such as patches, slots, Jerusalem crosses, loops, and meander lines have been explored to tailor the absorber response. Analytical, numerical, and experimental studies have contributed to a growing knowledge base, enabling reliable prediction and enhancement of FSS absorber performance under varying conditions.

1.3.1 Basic Theory and Concept

An FSS-based absorber typically consists of periodic metallic patterns printed on a dielectric substrate, with a continuous metallic ground plane (often copper) on the reverse side, as depicted in Figure 1.2. This ground plane eliminates transmission through the structure, ensuring that all incident energy is either reflected or absorbed. Given that the transmission coefficient $|S_{21}|$ is effectively zero, maximizing absorption requires minimizing the reflection coefficient $|S_{11}|$ at the desired frequencies.

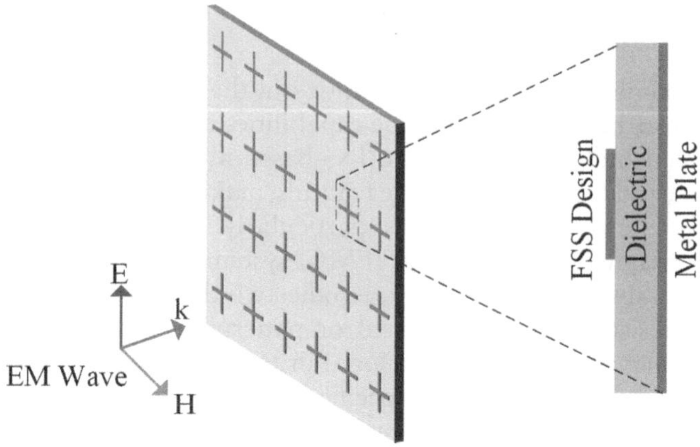

Figure 1.2 Typical schematic of a FSS absorber.

The absorptivity $A(\omega)$ of the structure is defined as:

$$A(\omega) = 1 - |S_{11}|^2 - |S_{21}|^2 \tag{1.1}$$

Since transmission is nullified, $S_{21} = 0$.

Therefore:

$$A(\omega) = 1 - |S_{11}|^2 \tag{1.2}$$

Here, $|S_{11}|^2$, $|S_{21}|^2$, and $A(\omega)$ represent the reflectance, transmittance, and absorptivity of the structure, respectively, at a given angular frequency ω. Thus, to achieve maximum absorption, the design goal is to minimize surface reflectance. The reflection coefficient at the surface is given as:

$$S_{11}(\omega) = \frac{Z_{in}(\omega) - Z_0}{Z_{in}(\omega) + Z_0} \tag{1.3}$$

where $Z_{in}(\omega)$ is the input impedance of the structure, and Z_0 is the intrinsic impedance of free space. From Equation (1.3), it is evident that reflection is minimized when $Z_{in} = Z_0$, i.e., when the structure is impedance-matched to free space. Consequently, the input impedance can be tuned by adjusting the geometry and material properties of the FSS to achieve optimal matching at specific frequencies, thereby enhancing absorption performance.

1.3.2 FSS Absorber Applications

FSS-based absorbers are widely deployed in stealth platforms to suppress backscattered radar signals. Their ability to be engineered with angular and polarization stability makes them particularly attractive for RCS reduction in airborne, naval, and ground-based military assets. Beyond stealth, FSS absorbers are also utilized in electromagnetic compatibility (EMC), electromagnetic interference (EMI) shielding, antenna isolation, and anechoic chamber design. In space and satellite technologies, such structures are explored for thermal and electromagnetic shielding. Their planar profile, mechanical flexibility, and compatibility with other RF/microwave systems enable seamless integration into conformal and layered composite designs.

1.3.3 Conventional Absorber versus FSS Absorber

Conventional absorbers such as carbon-based materials, Salisbury screens, Jaumann layers, and magnetic composites are typically bulkier and require significant thickness to achieve effective absorption. These structures also tend to be narrowband and are highly dependent on incidence angle and polarization. In contrast, FSS-based absorbers offer low-profile, lightweight solutions with the potential for multiband or wideband performance. Furthermore, FSS designs can be tuned for frequency agility and customized for specific incident wave characteristics through geometrical manipulation or the inclusion of active components. The trade-off, however, is the complexity involved in design, simulation, and fabrication.

1.3.4 Fabrication Techniques

The fabrication of FSS absorbers typically involves standard printed circuit board (PCB) technologies, photolithography, or screen printing, depending on the feature size and material selection. Conductive inks, carbon-loaded paints, and resistive films are often used to realize lossy elements. For flexible or conformal absorbers, substrates such as polyimide or silicone are utilized. Advanced techniques like inkjet printing, laser etching, and 3D printing are emerging as viable options for creating complex multilayered FSS structures with high precision. Accurate fabrication is essential to ensure the periodicity and symmetry required for predictable EM performance.

1.3.5 Experimental Testing

Experimental validation of FSS absorbers is critical to confirm simulation results and practical applicability. Measurements are typically carried out in anechoic chambers using free-space setups with standard gain horn antennas. Reflection, transmission, and absorption characteristics are evaluated under normal and oblique incidence, for both TE and TM polarizations. Time-domain gating may be employed to eliminate undesired reflections. In addition, techniques such as waveguide measurements and probe-based near-field scanning are used for compact samples or frequency-specific assessments. Ensuring repeatability and environmental stability is essential in the testing phase.

1.3.6 Limitations

Despite their many advantages, FSS-based absorbers are not without limitations. One major challenge is the sensitivity to fabrication tolerances, which can significantly affect performance due to the precise nature of resonant elements. Moreover, achieving broadband and angle-independent absorption simultaneously remains complex, especially in ultra-thin profiles. Another challenge is scalability; while FSS designs are efficient in simulations, scaling them to large apertures while maintaining uniformity can be difficult. Finally, integration with active or tunable components introduces design and reliability complexities that must be carefully addressed.

1.4 LITERATURE SURVEY ON FSS-BASED ABSORBERS

Based on extensive research reported in the literature, FSS-based absorbers can generally be classified into three categories: narrowband absorbers [44–58], bandwidth-enhanced absorbers [59–61], and broadband absorbers [62–65]. The most basic form of narrowband FSS absorbers exhibits strong absorption at a specific resonance frequency but offers limited bandwidth. To overcome this limitation, bandwidth-enhanced absorbers have been developed by employing multi-resonant unit cells or structural modifications, thereby broadening the effective absorption range. In contrast, broadband FSS absorbers are typically realized using the circuit analog absorber (CAA) concept, incorporating lumped resistive elements and multilayer geometries to achieve wideband performance. This section provides a concise review of the representative literature associated with each category in a sequential manner.

1.4.1 Narrowband FSS Absorbers

The first reported FSS-based absorber, introduced in 2006, was based on a split-ring resonator (SRR) configuration [44]. The structure consists of a periodic array of SRRs mounted on a resistive sheet with a thickness of 1 mm and a surface impedance of 377 Ω, corresponding to the intrinsic impedance of free space. To excite magnetic resonance, the incident magnetic field is oriented perpendicularly to the plane of the SRRs. This configuration results in narrowband absorption centered around 2 GHz. Despite its functional performance, the design suffers from practical limitations, including a non-planar profile and relatively large overall thickness, which may restrict its applicability in low-profile or conformal systems.

A significant advancement in narrowband absorber design was reported in 2008 [45], featuring a planar and low-profile structure. The absorber consists of an ELC resonator printed on the front side of a dielectric substrate, while a thin metallic strip is patterned on the back side. This configuration supports electric resonance, which drives the absorption mechanism. The structural and material parameters are carefully optimized to achieve impedance matching with free space at the target frequency, thereby minimizing both reflection and transmission. The design achieves a strong single-band absorption peak of 96% at 11.65 GHz. Compared to the earlier SRR-based design in [44], this structure offers the advantages of being both thin and planar, making it more suitable for practical integration.

Another FSS absorber based on an electric-field-coupled ELC resonator was reported in [46], featuring a fully copper-laminated back side that serves as a ground plane. The basic ELC unit cell demonstrates a single-band narrowband absorption centered at 9.92 GHz. To achieve multiband absorption, the ELC resonators are arranged in arrays with a geometric scaling factor denoted by k_1. A 2 × 2 array configuration produces dual-band absorption at 9.18 GHz and 12.36 GHz. Expanding to a 3 × 3 array results in triple-band absorption at 7.44 GHz, 10.37 GHz, and 12.27 GHz. This demonstrates the scalability and flexibility of ELC-based designs for realizing multi-resonant absorption behavior through array arrangement and geometric tuning.

A conformal, polarization-insensitive triple-band absorber based on a modified triple circular ring resonator was presented in [47]. The design consists of three concentric circular ring resonators printed on a thin GML 1000 substrate with a thickness of 0.502 mm. A continuous metallic layer is laminated on the back side to eliminate transmission. This structure achieves triple-band narrowband absorption with resonant peaks at 4.19 GHz, 6.64 GHz, and 9.95 GHz. The conformal and polarization-insensitive nature of the design makes it suitable for versatile electromagnetic absorption applications.

1.4.2 Bandwidth-Enhanced FSS Absorbers

Several techniques have been proposed to enhance the bandwidth of FSS-based absorbers [59–61]. One of the conventional approaches involves integrating multiple resonant elements within a single unit cell. By carefully selecting the geometrical parameters, the resonance frequencies of these elements can be brought closer together, resulting in a broader absorption bandwidth. A report design in [59] illustrates one such bandwidth-enhanced absorber design and its corresponding absorption response. In this structure, the unit cell comprises two ELC resonators of identical shape but differing dimensions, arranged in a 2 × 2 array configuration. This arrangement enables the excitation of closely spaced resonant modes, thereby widening the overall absorption bandwidth.

Although using an array of resonant structures can enhance bandwidth, it is often limited by the need for large unit cell dimensions. To overcome this limitation and achieve bandwidth enhancement without increasing the unit cell size, multi-resonating structures are utilized. One such absorber was reported in [60], where a hexagonal interdigital resonator is designed to support dual resonance. By tuning the resonant frequencies of the two modes to be closer together, a bandwidth-enhanced absorption response is achieved.

Other techniques for enhancing bandwidth involve the use of multilayer geometries. Reported work in [61] illustrates one such bandwidth-enhanced absorber, where six alternating metal-dielectric layers are stacked together. In this configuration, the absorber achieves an absorptivity of over 80% within the frequency range of 8.8–10.8 GHz, corresponding to a relative full width at half maximum (FWHM) absorption bandwidth of 23%.

1.4.3 Broadband FSS Absorbers

While bandwidth-enhanced absorbers typically offer bandwidths limited to a few hundred MHz, practical applications often demand much wider absorption ranges. To meet this requirement, several broadband absorbers have been developed, achieving absorption bandwidths on the order of several GHz. One widely adopted approach is the circuit analog absorber (CAA), where lumped resistors are integrated across gaps in the metallic pattern on a dielectric substrate. These broadband absorbers generally consist of a two-layer configuration, with a resistive surface placed above a metallic ground plane separated by a dielectric spacer. An early example of such a design is reported in [62], which achieves a 1.5 GHz bandwidth under normal incidence, maintaining absorptivity above 90%. The use of high-ohmic-value lumped resistors enables effective absorption over a broad frequency range. Additionally, this absorber demonstrates polarization insensitivity and strong angular stability, making it suitable for diverse real-world applications.

To further enhance the bandwidth of FSS-based absorbers, the design reported in [63] integrates the principles of multiple resonators and multilayer architecture. The structure consists of a multilayer configuration where the top layer features resistor-loaded double square loops, while the bottom layer includes resistor-loaded single square loops. The underside of the bottom layer is fully metal-backed, acting as a ground plane. This hybrid design enables ultra-wideband absorption, achieving a remarkable fractional bandwidth of 114.40%, covering the frequency range 4.96–18.22 GHz.

Another broadband absorber design was reported in [64], where the capacitive reactance inherent to the square loop geometry is strategically employed to reduce the overall thickness of the structure. The top layer consists of multiple pairs of square loop resonators interconnected via lumped resistors. This configuration enables a broad absorption bandwidth of 100.26%, covering the frequency range from 1.91 GHz to 5.75 GHz. The use of capacitive coupling not only contributes to bandwidth enhancement but also facilitates a low-profile absorber design.

A dual-polarized ultra-wideband absorber was presented in [65], utilizing multimode resonances generated by four rotationally symmetric bent metallic strips. These metallic strips, integrated with lumped resistors and printed on a lossy dielectric layer, support three distinct resonant modes. In each mode, current flows through different pairs of chip resistors, leading to effective energy dissipation. The design achieves polarization insensitivity due to its rotational symmetry and provides an ultra-wideband absorption bandwidth of 127.9%, spanning from 2.68 GHz to 12.19 GHz.

1.5 FREQUENCY SELECTIVE RASORBER

FSS-based absorbers have been extensively explored in literature due to their critical role in stealth technology and electromagnetic shielding. However, a major limitation of conventional absorbers is that they completely block communication signals, making them unsuitable for systems requiring simultaneous signal transmission and absorption [66, 67]. To address this issue, a novel class of structures has been developed by integrating a conventional FSS absorber with a bandpass-type FSS. These hybrid structures are referred to as frequency selective rasorbers (FSRs), absorptive frequency selective transmission (AFST) structures, or simply rasorbers [68]. The term "rasorber" originates from the fusion of the words "radome" and "absorber," indicating a device capable of both signal transparency and energy absorption.

FSRs have gained substantial attention in recent years for their unique ability to simultaneously provide selective transmission within a designated passband and absorption outside it. A general schematic of an FSR design is shown in Figure 1.3. Typically, the structure consists of a resistive layer and a bandpass FSS layer. The resistive layer is responsible for absorbing incident energy outside the passband, while the bandpass layer ensures efficient transmission within the desired frequency range. Importantly, the integration of these layers must follow a carefully designed methodology to maintain spectral selectivity ensuring that the transmission band remains unaffected by the absorber's dissipative behavior.

FSRs have proven highly effective in stealth applications, where the goal is to reduce the detectability of a target by absorbing unwanted signals while allowing essential communication signals to pass through. In such systems, minimizing the RCS (measure of how detectable an object is by radar) is critical. By incorporating a transmission band aligned with the operating frequency of a radiating system, FSRs can function as radomes that ensure low RCS outside the passband. Numerous studies in literature have demonstrated the successful implementation of such low-RCS systems. FSRs are also referred to as absorptive frequency selective transmission (AFST) structures. In contrast, structures that reflect signals in a specific band while absorbing outside it are termed absorptive frequency selective reflection

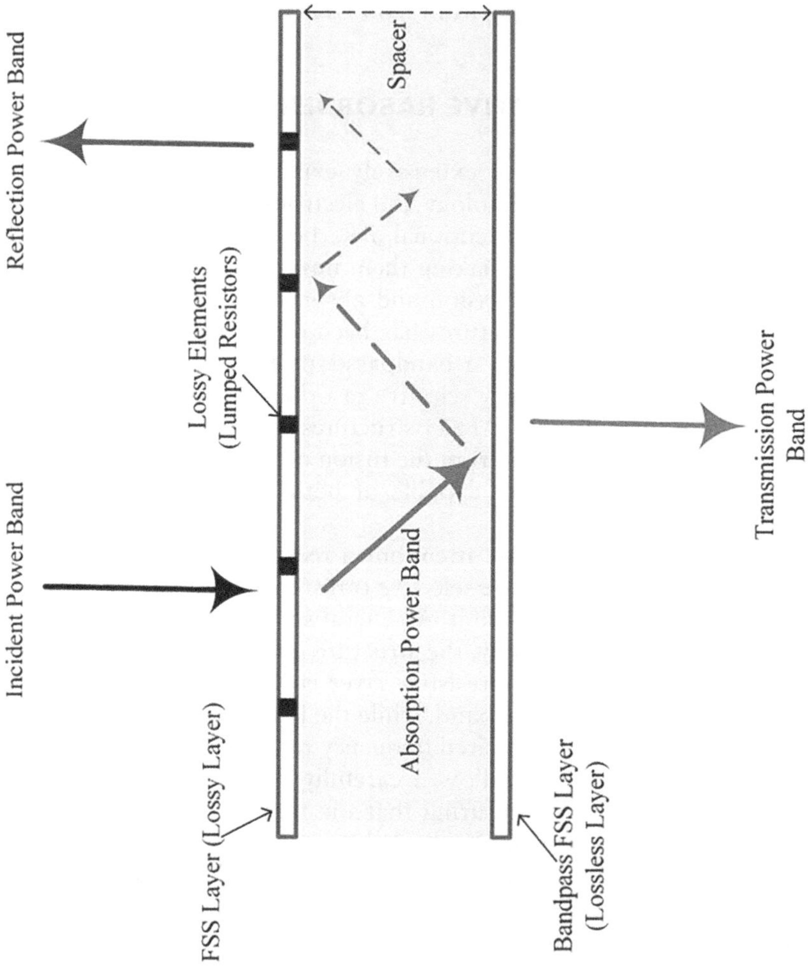

Figure 1.3 Schematic representation depicting the working of the general frequency selective rasorber.

(AFSR) structures [69]. In AFSR designs, a reflection band is deliberately introduced alongside one or more absorption bands [70–75]. AFSRs are often employed as substrates for antennas or radiating systems where the operational frequency aligns with the reflection band, enabling effective radiation while suppressing out-of-band scattering. This configuration is particularly useful for achieving low RCS profiles without compromising the antenna's performance.

1.5.1 Design Concept and Objectives

A typical FSR structure comprises two key functional layers: a resistive layer and a bandpass FSS. Each layer plays a distinct yet complementary role in controlling the EM wave interaction with the structure. The resistive layer is responsible for absorbing incident EM energy that lies outside the desired transmission band. This absorption is achieved by converting the incoming EM energy into losses using resistive components, commonly resistive sheets or lumped resistors integrated into a patterned metallic surface. As a result, the resistive layer significantly reduces unwanted reflections and prevents the transmission of out-of-band signals, effectively acting as a wideband absorber for non-communicative frequencies.

The bandpass FSS layer, in contrast, is engineered to selectively allow transmission within a specific frequency range. This layer consists of carefully designed metallic patterns (such as slots, loops, or patches) that resonate at the desired transmission frequency, enabling EM waves within that range to pass through with minimal attenuation. Signals outside this passband are inherently rejected or attenuated due to the filter-like behavior of the FSS.

Crucially, integrating these two layers into a single FSR structure requires precise electromagnetic and structural design. The primary challenge lies in ensuring that the lossy resistive layer does not interfere with the selective transmission characteristics of the bandpass FSS. To accomplish this, the layers must be carefully aligned and often separated by an optimized dielectric or air gap. Transmission pole resonators may be designed and incorporated within the lossy layer for maintaining the overall passband. Parameters such as material properties, geometric layout, and inter-layer spacing must be fine-tuned to achieve a balance between absorption and transmission performance.

The fundamental objective of an FSR is thus twofold:

1. Maximize absorption of unwanted signals outside the passband to achieve low radar cross-section (RCS) or effective shielding.
2. Maintain low insertion loss and high transmission efficiency within the passband to support unhindered communication or sensing.

This dual capability makes FSRs particularly valuable in stealth technology, electromagnetic compatibility (EMC), and advanced antenna systems where both transparency and concealment are required within the same structure.

1.5.2 Classifications and Significance

FSRs can be systematically categorized based on their spectral characteristics and intended application requirements. The classification helps in understanding the trade-offs between absorption bandwidth, structural complexity, thickness, and transmission performance. Accordingly, FSRs can be classified as narrowband and broadband types.

1.5.2.1 Narrowband Rasorber

Narrowband rasorbers are specialized electromagnetic structures designed to exhibit strong absorption at a specific, narrow frequency range that is typically centered around a single resonant frequency. These structures rely primarily on resonant mechanisms provided by carefully designed FSS. Because the absorption mechanism is based on resonance rather than broadband dissipation, no additional lossy materials such as resistive inks or lumped resistors are typically required. The narrowband behavior is achieved through FSS geometries that induce destructive interference and impedance matching at the resonant frequency, minimizing both reflection and transmission. These rasorbers are usually physically thin, lightweight, and relatively simple to fabricate. As a result, they are ideal for applications that require selective filtering or suppression of a narrow band of unwanted signals such as in radar jamming, notch filtering, or electromagnetic compatibility in communication systems. However, the main limitation of narrowband rasorbers is their lack of flexibility in covering wide spectral regions.

1.5.2.2 Broadband Rasorber

Broadband rasorbers are designed to absorb electromagnetic waves over a wide frequency range, typically covering multiple GHz bands. These structures are more sophisticated and are intended for advanced stealth and EMI suppression applications where suppression of a broad range of frequencies is essential. To achieve such broadband performance, these FSRs typically incorporate lossy elements, including lumped resistors, resistive films, or resistive surface patterns. These components are integrated into or placed over the FSS layer to dissipate energy over a wider frequency range. The combination of a resistive layer and a bandpass FSS ensures that while the unwanted out-of-band signals are absorbed, the desired in-band signals pass through with low insertion loss. Broadband FSRs often require multilayer designs, optimized impedance matching, and precise material selection to maintain effectiveness across the intended spectrum. They are especially well suited for stealth platforms, such as aircraft, UAVs, and ground vehicles, where RCS must be minimized over a large bandwidth without degrading communication or sensing capability. While more complex and sometimes thicker than their narrowband counterparts, broadband rasorbers provide a more robust solution for real-world operational environments involving dynamic and multifrequency threats.

1.5.3 Applications

The unique ability of FSRs to combine electromagnetic wave absorption with selective transmission makes them highly valuable in a broad spectrum of applications. These range from military stealth systems to civilian electronic environments, where precise control over the interaction between materials and electromagnetic waves is critical.

1.5.3.1 Low Radar Cross-Section Radiating Systems

One of the most impactful applications of FSRs lies in the domain of stealth technology and radar cross section (RCS) management. In military platforms such as aircraft, ships, missiles, and ground vehicles, minimizing RCS is essential for evading radar detection. However, these platforms must also maintain critical communication, navigation, and sensing functionalities. FSRs provide an innovative solution by acting as intelligent electromagnetic windows. They are designed to absorb incoming radar signals outside a narrow operational frequency band while simultaneously allowing transmission of signals within that band. This enables the integration of antennas and sensors into stealth platforms without compromising their low-observability characteristics. For example, in airborne systems, an FSR can serve as a radome material that transmits communication or surveillance signals (e.g., in X or Ku band), while absorbing hostile radar signals (e.g., in S or C band). This results in reduced detectability without degrading system performance. The incorporation of FSRs into antenna design also allows for low-RCS antenna systems, where the antenna's reflecting surface or backing structure contributes to the overall stealth profile of the platform.

1.5.3.2 Electromagnetic Shielding and Interference Reduction

Electromagnetic compatibility (EMC) and electromagnetic interference (EMI) control are critical challenges in modern electronics, especially in dense environments such as hospitals, research labs, avionics, and military installations. Unwanted electromagnetic emissions can cause signal distortion, data loss, or malfunction of sensitive electronic systems. FSRs offer a frequency selective shielding mechanism, allowing essential signals (e.g., for communication or sensing) to pass through, while suppressing disruptive or spurious signals. This selective filtering helps maintain clean signal environments, reducing the risk of interference in complex, multi-device electronic systems. Additionally, FSRs are useful in shielded enclosures and EM-safe equipment housing, where conventional shielding may be too restrictive or bulky. By engineering the passband and absorption characteristics, FSRs provide a customizable and compact solution for shielding without the need for additional hardware-based filters or external circuitry.

1.5.3.3 Other Applications

Beyond their use in stealth and EMI control, FSRs are being explored for a variety of innovative applications across both defense and civilian sectors:

- *Satellite Communication*: In satellite payloads and ground terminals, FSRs can be used to isolate uplink and downlink frequencies, minimize cross-talk, and improve antenna isolation – all while keeping the form factor compact.
- *Medical Imaging Systems*: In systems like MRI and other high-frequency imaging modalities, FSRs can be used to shield diagnostic equipment from ambient electromagnetic noise, ensuring high-resolution image quality and patient safety.
- *Wireless Sensing and IoT*: FSRs enable reconfigurable environments for smart sensing, where only specific frequency bands are allowed for transmission, enhancing signal fidelity, battery efficiency, and security in sensor networks.
- *High-Frequency Test and Measurement Systems*: In RF/microwave laboratories and clean measurement environments, FSRs are valuable in isolating experimental setups from broadband interference while still allowing instrumentation signals to pass.
- *Reconfigurable RF Systems*: With the integration of tunable or switchable elements such as varactors, MEMS, or graphene-based materials, FSRs are paving the way for adaptive and reconfigurable systems that can change their electromagnetic response dynamically in real time.

1.6 LITERATURE SURVEY ON FREQUENCY SELECTIVE RASORBERS

In recent years, the design and development of frequency selective rasorbers (FSRs) have garnered considerable attention due to their versatile electromagnetic properties and practical applications. As research in this area has advanced, numerous efforts have focused on enhancing the performance metrics of FSRs, such as absorption bandwidth, transmission efficiency, angular stability, and polarization insensitivity. Based on their spectral response characteristics, FSRs are broadly categorized into several distinct types. These include structures with a lower transmission band followed by an upper absorption band (T-A) [76–81], those with a lower absorption band and an upper transmission band (A-T) [82–90], designs featuring a transmission band flanked by two absorption bands (A-T-A) [91–107], and configurations with alternating multiple absorption and transmission bands (A-T-A-T) [108–110]. Additional hybrid variants have also been proposed in the literature [111]. This section provides a sequential overview of recent FSR designs corresponding to these classifications, highlighting their structural features, operational mechanisms, and performance outcomes.

1.6.1 T-A FSR

One of the earliest demonstrations of an FSR exhibiting a lower transmission band followed by an upper absorption band (T-A type) was reported in [76]. The structure comprises a resistive frequency selective surface (FSS) layer positioned above a bandpass FSS-based on interdigitated Jerusalem cross elements. This configuration enables a distinct transmission band centered at 4.6 GHz, along with a wide absorption band spanning from 10 GHz to 18 GHz. Building upon this concept, several other T-A-type FSRs have been developed [77–81]. For example, a dual-polarized T-A FSR reported in [77] incorporates meander-line square loop elements on both the lossy and lossless layers. The lossy layer includes lumped resistors integrated within the loops, which contribute to the absorption mechanism. This design achieves an effective absorption band of 4.8–6.8 GHz, while remaining largely transparent to incident waves below 1.54 GHz.

1.6.2 A-T FSR

Several FSR designs featuring a lower absorption band followed by an upper transmission band (A-T type) have been reported in the literature [82–90]. One such example is the A-T FSR presented in [82], which is implemented by integrating a parallel LC resonator onto a resistive surface. This configuration achieves a wide absorption band spanning 3–9 GHz, while maintaining a distinct transmission band centered at 10.2 GHz.

The rasorber design presented in [83] features dipole-like elements on the resistive layer and slot arrays on the lossless layer. The dipole-like structures are printed on both sides of the resistive substrate and interconnected using vias, with lumped chip resistors mounted between them. This configuration provides a transmission band centered at 5.6 GHz, while achieving lower-band absorption of 2.8–5 GHz. Although the design uses only two lumped resistors per unit cell, making it relatively simple, it exhibits polarization sensitivity due to its asymmetrical geometry.

A compact FSR design presented in [84] demonstrates a wide transmission band ranging from 8.3 GHz to 11.07 GHz, positioned above an absorption band extending from 2.4 to 7.1 GHz. The resistive layer incorporates a high-inductance circular spiral resonator, whose parallel resonance facilitates the broad upper transmission band. Meanwhile, the bandpass layer employs a triple-layer hexagonal aperture configuration to support the desired frequency response.

In the work reported in [85], an FSR structure is designed using a centrosymmetric bent-strip resonator. This resonator is tuned to resonate at the transmission band of the underlying bandpass FSS, resulting in an upper passband. The A-T rasorber achieves a transmission peak at 12.76 GHz, with the absorption band extending from 6.10 GHz to 10.98 GHz.

1.6.3 A-T-A FSR

Among the various categories of FSRs, the A-T-A type is particularly challenging to design due to the potential overlap between transmission and absorption bands, which can significantly degrade performance within the transmission band. Nonetheless, several A-T-A FSR designs have been proposed in recent years [91–107]. The design presented in [91] consists of a two-layer structure comprising a resistive layer and a bandpass layer. A notch resonator printed on the back side of the resistive layer introduces a transmission pole aligned with the bandpass layer. This rasorber exhibits a clear transmission window centered at 5.7 GHz, flanked by absorption bands extending from 2.6 GHz to 5.2 GHz (lower) and 6.0 GHz to 8.5 GHz (upper).

To achieve a broader absorption bandwidth, a rasorber employing a dual lossy-layer configuration has been proposed in [92]. In this design, each lossy layer contributes an independent absorption band. When only one lossy layer is used, the structure exhibits two distinct absorption bands spanning 2.3–4.5 GHz and 7.9–10.3 GHz, with a transmission window centered at 6.3 GHz. However, incorporating both lossy layers significantly broadens the absorption range to 2.3–5.3 GHz and 7.8–14.6 GHz, while maintaining the same transmission frequency.

An FSR designed using the equivalent circuit model is presented in [93]. The structure features a lossy layer with a Jerusalem element and a parallel resonance element at the center, while the lossless layer consists of a multilayer bandpass FSS to achieve a highly selective and wide transmission band. The transmission band spans from 7.9 GHz to 9.0 GHz, while the two adjacent absorption bands cover 3.3 GHz to 7.1 GHz and 9.4 GHz to 12.0 GHz.

To meet the needs of systems operating at two different frequencies for two polarizations, an FSR has been reported in [94] that exhibits anisotropic transmission bands for each polarization. The FSR consists of a split circular resonator with metallic strips on the top resistive layer, while the bottom bandpass layer incorporates two types of slots corresponding to the two polarizations. The rasorber achieves a transmission at 10.40 GHz with absorption bands between 6.26 GHz and 9.58 GHz and 11.46 GHz ad 14.06 GHz for the TE incident wave. For the TM incident wave, the transmission shifts to 11.40 GHz, and the absorption occurs within the frequency ranges of 6.35 GHz to 10.42 GHz and 12.69 GHz to 14.64 GHz.

A recent A-T-A FSR design based on resistive cross dipoles coupled with electric field resonators has been reported in [95]. The Rasorber is fabricated on a single 6.5 mm thick substrate, with the bottom bandpass layer printed on the opposite side. The FSR exhibits transmission at 6.01 GHz, with absorption bands ranging from 2.95 GHz to 5.65 GHz and 7.4 GHz to 9.24 GHz. Additionally, the design demonstrates angular stability up to 50°.

1.7 A-T-A-T FSR

With ongoing advancements in electromagnetic structures, significant research has been directed toward the development of dual-band transmissive frequency selective rasorbers (FSRs). Several designs featuring alternating absorption and transmission (A-T-A-T) characteristics have been documented in the literature [108–110]. For instance, [108] presents an FSR that achieves dual transmission bands using a lumped-resistor-loaded metallic dipole with a dual-resonance configuration embedded within the center of the resistive layer. The frequency selective surface (FSS) employed for bandpass operation consists of two adjacent slots with differing geometries. This configuration achieves transmission windows at 7.7 GHz and 12.6 GHz, while providing absorption bands spanning 4.02–6.27 GHz and 8.21–12.02 GHz, respectively.

In [109], a dual-polarized FSR with a hexagonal unit cell is introduced. The lossy layer features a central loop loaded with resistors, supplemented by three folded loops that generate three transmission poles. This structure supports transmission at 6.1 GHz and 10.1 GHz, with absorption occurring over the ranges of 3.30–4.97 GHz and 7.42–9.03 GHz.

Further development is presented in [110], where an A-T-A-T-type FSR capable of independently controlled dual-band transmission is proposed. The lossy layer is composed of four square rings and two stubs, where each stub's length can be adjusted independently to tune the two transmission poles. This design achieves passbands at 8.0 GHz and 11.9 GHz, within a broadband −10 dB reflection range extending from 5.0 GHz to 12.8 GHz.

1.8 RCS-REDUCTION-BASED ON FSR

In recent years, considerable research has focused on the development of radiating systems with reduced radar cross section (RCS), driven by the growing demand for stealth technology. RCS is a critical parameter that quantifies how detectable a target is by radar; minimizing this value is essential for improving the concealment and survivability of platforms in defense applications. Consequently, reducing the RCS of radiating systems used in stealth platforms is of paramount importance. Various strategies have been proposed in the literature to achieve low-RCS radiators [112–136]. These include techniques such as geometrical shaping of antennas [112–115], application of radar absorbing materials (RAM) [116–121], use of polarization conversion structures [122, 124], and integration of artificial magnetic conductor (AMC) surfaces [125, 126]. However, many of these methods suffer from inherent limitations, including compromised radiation performance, narrowband RCS reduction, and sensitivity to incident angle.

A notable approach is presented in [127], where a low-RCS antenna is realized by integrating it with an anisotropic frequency selective transmission (AFST) structure. The AFST structure, is composed of paired circular slot resonators and metallic strips. It facilitates a transmission band centered at 8.9 GHz for vertical polarization, while exhibiting absorption in the horizontal polarization. A patch antenna, designed to operate at 8.9 GHz and placed beneath the AFST layer, demonstrates more than a 10 dB RCS reduction in both polarizations outside its operational band.

In another study [128], an antenna is integrated with an anisotropic frequency selective rasorber (AFSR) that functions as a band-notch absorber. This AFSR structure generates a reflection notch under vertical polarization and exhibits broadband absorption in horizontal polarization. A dipole antenna, whose operating frequency aligns with the reflection notch of the AFSR, is mounted atop the structure. The resulting integrated system achieves more than 10 dB RCS reduction in both polarizations outside the antenna's passband.

Further advancement is reported in [129], where a three-dimensional AFSR structure is developed, featuring a reflection notch flanked by absorption bands. A monopole antenna mounted over this 3D AFSR configuration demonstrates significant RCS suppression outside its operating frequency, with −10 dB RCS reduction bandwidths of 64.7% and 41% in the respective polarizations.

1.9 PRACTICAL CHALLENGES

Despite their promising features, the design and implementation of frequency selective rasorber (FSR) present several practical challenges that can impact performance and scalability.

1.9.1 Material Choice and Availability

The performance of FSR is highly dependent on the properties of the materials used, especially the resistive and dielectric layers. Material parameters such as sheet resistance, permittivity, and loss tangent must be carefully chosen and consistently maintained during fabrication. However, the limited availability of advanced low-loss and temperature-stable materials often restricts design flexibility. Furthermore, commercial substrates may not support ultra-thin conductive or resistive films required for fine-tuned frequency responses, leading to compromises in performance.

1.9.2 Manufacturing Challenges

FSR structures involve multilayer assemblies with sub-wavelength geometric precision. Achieving precise alignment between resistive sheets and FSS layers can be technically demanding. Additionally, fabrication tolerances, such as etching accuracy and film deposition uniformity, play a critical role in the final spectral response. Manufacturing inconsistencies may introduce deviations from the simulated performance, particularly at higher frequencies where even small dimensional errors have significant impact.

1.9.3 Potential Solutions

To address these challenges, several strategies can be adopted. Advanced lithographic techniques and precision screen printing can help improve fabrication accuracy. Emerging additive manufacturing and inkjet printing methods offer a cost-effective and scalable approach for producing complex FSR geometries. On the material front, research into novel metamaterials, graphene-based conductors, and tunable substrates is paving the way for enhanced functionality and environmental robustness. Moreover, implementing tunable or reconfigurable elements within the FSR can provide adaptability across varying conditions, reducing the sensitivity to fabrication tolerances and environmental shifts.

1.10 MOTIVATION AND SCOPE OF BOOK

In today's rapidly evolving world, advances in communication technologies continue to accelerate. At the core of wireless communication systems lies electromagnetic (EM) wave propagation. With increasing technological demands, precise control over EM wave propagation has emerged as a prominent area of research. Among the various strategies, EM wave absorption has become critically important in defense applications, particularly for reducing radar cross-section (RCS) to enhance target concealment. In addition to stealth, EM absorption is also vital in electromagnetic interference (EMI) mitigation and shielding applications.

Frequency selective surface (FSS)-based absorbers offer unique advantages over conventional absorbers due to their thin profile, lightweight structure, and inherent frequency selectivity. These characteristics make them especially attractive for stealth applications. Although FSS absorbers are typically narrowband, broadband absorption can be achieved through the use of multilayer circuit analog designs incorporating lossy components.

However, FSS absorbers inherently block communication signals at all frequencies. This limitation has spurred growing interest in frequency selective rasorbers (FSRs), also referred to as anisotropic frequency selective transmission (AFST) structures. These structures integrate the bandpass functionality of FSS with absorptive behavior, allowing selective transmission and absorption across different frequency bands depending on the application requirements.

While numerous single-layer FSS absorbers achieving single or multiple narrowband absorptions have been reported [44–58], literature on single-layer rasorbers with narrowband absorption is relatively scarce. Most existing FSRs employ multilayer configurations with broadband absorption. Therefore, an opportunity exists to evolve single-layer, multiband absorbers into narrowband rasorbers. A key objective of this book is to understand the development of a thin, single-layer rasorber capable of supporting multiple narrowband absorption and transmission bands. The early part of the book focuses on designing flexible, conformal FSR structures with multiple absorption and transmission bands.

Typically, FSRs are categorized based on the sequence of their absorption and transmission bands, including configurations such as T-A, A-T, T-A-T, A-T-A, and A-T-A-T. Recent research has explored various FSR configurations to improve performance. For example, the A-T structures reported in [83, 85] use asymmetric unit cells, making them polarization-sensitive. Conversely, the dual-polarized designs in [84, 86] exhibit wider absorption bandwidths but require four structural layers and a high number of lumped elements per unit cell (6 and 8, respectively). This highlights a need for improved A-T FSRs that combine broad absorption, reduced component count, polarization insensitivity, and simpler construction. This book presents the designs of such an improved A-T rasorber design. It also explores a novel T-A-T configuration presenting a comprehensive design methodology for its realization.

Among the various configurations, A-T-A rasorbers, which feature a transmission band flanked by two absorption bands, are highly suitable for applications such as out-of-band RCS and EMI reduction. However, these designs are challenging due to the spectral proximity of the absorption and transmission bands, which often results in degraded insertion loss within the passband. For effective interference mitigation and shielding in narrowband systems, high selectivity characterized by a steep transmission notch and high quality factor (Q) is essential. Existing A-T-A FSRs [91–107] generally lack this sharp selectivity, making them less suitable for high-performance narrowband applications. This book presents a compact, polarization-insensitive A-T-A FSR design with a high-Q transmission band, intended for use in low-RCS radome applications. The book further presents a miniaturized dual-band A-T-A-T FSR configuration, with improved performance over the reported designs [108–110] in terms of absorption bandwidth, polarization stability, compactness, and angular independence.

Finally, absorptive frequency selective reflection (AFSR) structures representing another important category is also provided in this book. These structures function as band-notched absorbers, combining a reflection band with one- or two-sided absorption bands. AFSRs can serve as ground planes for antennas, enabling out-of-band RCS suppression [69]. Several AFSR designs have been explored in the literature [70–75]. For instance, [128] investigates a dipole antenna integrated with a single-polarization AFSR to achieve RCS reduction. Further studies in [129] examine monopole and dipole antennas mounted on 3D AFSR structures for low-RCS operation. Notably, a polarization-insensitive AFSR is particularly well suited for dual-polarized systems. This book presents an integration strategy for a dual-polarized crossed dipole antenna with a polarization-insensitive AFSR structure to achieve efficient RCS reduction.

Polarization-insensitive rasorber/absorber designs presented in this book are:

- Narrowband flat/flexible FSR on a single-layer substrate.
- A-T/T-A-T rasorber with improved performance.
- Highly selective A-T-A FSR.
- Low-RCS radiation system co-designed with an A-T-A FSR.
- Miniaturized A-T-A-T FSR.
- Low RCS dual-polarized radiation system co-designed with AFSR structure.

1.11 ORGANIZATION OF BOOK

This book is organized into eight chapters, beginning with an overview of frequency selective rasorber (FSR) technology in Chapter 1. The chapter introduces the fundamental concept of electromagnetic (EM) wave absorption and traces the evolution of frequency selective surfaces (FSS) and FSRs, highlighting key developments and current state-of-the-art designs. The motivation for the research and a summary of the main contributions are also outlined to provide context for the work that follows.

Chapter 2 presents a novel design of a T-A-T FSR that integrates a split-ring absorber and a dual-band bandpass FSS on opposite sides of a single-layer substrate with a thickness of 0.8 mm. This structure achieves narrowband resonant absorption between two well-defined transmission bands, demonstrating the feasibility of a compact, dual-band transmission FSR using a single substrate layer.

In Chapter 3, the focus shifts to flexible FSR designs, specifically A-T-T-A (2T2A) and A-T-A-T-A (2T3A) configurations. These rasorbers exhibit dual or triple absorption bands in conjunction with dual-band transmission. Constructed on a flexible substrate, they combine multiband absorptive layers with dual-bandpass FSS layers. The chapter also investigates the conformal behavior of these designs under bending angles of 120°, 150°, and 180°, confirming their suitability for integration in curved and flexible platforms.

Chapter 4 introduces an A-T FSR that features a higher-frequency transmission band positioned beyond a lower broadband absorption region. The design is achieved by first developing separate broadband absorber and bandpass FSS structures, which are subsequently integrated to preserve both functionalities in a single device.

A compact T-A-T FSR is proposed in Chapter 5, where dual transmission bands are located on either side of a central broadband absorption band. This two-layer structure combines a Jerusalem-cross-based dual-band FSS with a resistive square loop absorber. The integration ensures simultaneous realization of both absorption and transmission features within a compact footprint.

Chapter 6 proposes an A-T-A FSR that offers high passband selectivity, a key feature for effective electromagnetic shielding and radar signature management. The design utilizes a high-Q Minkowski fractal resonator printed on the resistive layer to establish a sharp transmission pole, aligned with the passband of a double cross-slot-shaped FSS. Additionally, the chapter demonstrates how the proposed FSR can be integrated with a patch antenna to create a low-RCS radiating system, achieving significant monostatic and bistatic RCS reduction while maintaining desirable antenna performance.

In Chapter 7, a novel miniaturized A-T-A-T FSR is introduced using a compact design strategy. The resulting miniaturized structure effectively achieves two distinct transmission windows separated by absorption bands, offering enhanced spectral control for multifunctional EM applications.

In the last Chapter 8, a novel strategy for realizing low-RCS dual-polarized antenna is presented. A polarization-insensitive, compact absorptive frequency selective reflection (AFSR) structure is developed and integrated with a dual-polarized crossed dipole antenna. This integration demonstrates notable monostatic and bistatic RCS suppression compared to traditional reflector-backed configurations, making it a promising solution for stealth communication systems.

1.12 OUTCOME OF BOOK

The chapterwise outcomes of FSR designs presented in this book are summarized in Table 1.1.

Table 1.1 Summary of Proposed Polarization-Insensitive Rasorber/Absorber Designs

Absorption	*Chapter No.*	*Rasorber Type*	*Achieved Parameters*
Narrowband	2	T-A-T	Single layer; Absorption: 6.2 GHz; B.W. = 110 MHz (1.75%) Transmission 1: 3.1 GHz; I.L. = 0.6 dB Transmission 2: 8.4 GHz; I.L. = 1.2 dB
	3	A-T-T-A (2T2A), A-T-A-T-A (2T3A)	Single Layer; Flexible; Conformal response tested up to 180° curvature angle. 2T2A: A− 6.4 and 14.7 GHz; T− 9.6 and 13.4 GHz 2T3A: A-6.6, 11.1, and 14.7 GHz: T− 9.6 and 13.4 GHz
Broadband	4	A-T	Absorption: 3.1–8.7 GHz (94.91%) Transmission: 11.4 GHz; I.L. = 0.7 dB
	5	T-A-T	Absorption: 3.6–7.4 GHz (69.09%) Transmission 1: 2.1 GHz; I.L. = 1.4 dB Transmission 2: 8.8 GHz; I.L. = 1.5 dB
	6	A-T-A	Absorption 1: 3.4–7.5 GHz (75.22%) 80% Absorption 2: 9.2–10.8 GHz; 16% Transmission: 8.4 GHz; I.L. – 0.7 dB FSR + patch antenna: Average monostatic RCS reduction = 11.92 dB (lower band) and 5.04 dB (upper band) Average bistatic RCS reduction = 77% (lower band) and 60% (upper band)
	7	A-T-A-T	Absorption 1: 3.7–10.5 GHz (95.71%) Absorption 2: 12.6–14.6 GHz; (14.70%) Transmission 1: 11.7 GHz; I.L. = 1.0 dB Transmission 2: 16.0 GHz; I.L. = 0.7 dB
	8	AFSR	Absorption 1: 4.2–7.0 GHz (50%) Absorption 2: 9.2–11.5 GHz (22.2%) Reflection: 8.2 GHz attenuation = 1.5 dB AFSR + crossed dipole: Average monostatic RCS reduction = 12 dB Average bistatic RCS reduction = 80%

Note: A: absorption; T: transmission; B.W.: bandwidth; I.L.: insertion loss

REFERENCES

1. T. K. Wu, "Frequency Selective Surface and Grid Array," 1st ed. New York, NY, USA: Wiley, 1995.
2. W. F. Bahret, "The Beginnings of Stealth Technology," IEEE Trans. Aerosp. Electron. Syst., vol. 29, no. 4, pp. 1377–1385, Oct. 1993.
3. B. A. Munk, "Frequency Selective Surfaces: Theory and Design," 1st ed. New York, NY, USA: Wiley, 2000.
4. S. W. Lee, "Scattering by Dielectric-Loaded Screen," IEEE Trans. Antenn. Propag., vol. 19, pp. 656–665, 1971.
5. E. Pelton and B. Munk, "A Streamlined Metallic Radome," IEEE Trans. Antenn. Propag., vol. 22, no. 6, pp. 799–803, Nov. 1974.
6. X. Yuan and X. Yuan, "A Transmissive/Absorbing Radome with Double Absorbing Band," Microwave Opt Technol Lett., vol. 58, no. 8, pp. 2016–2019, Aug. 2016.
7. S. Agahi and R. Mittra, "Design of a Cascaded Frequency Selective Surface as a Dichroic Subreflector," In Proceedings of the Antennas and Propagation Society International Symposium: AP-S Merging Technologies for the 90' (Digest)., Dallas, TX, USA, 7–11, pp. 88–91, May 1990.
8. F. Costa, S. Genovesi, A. Monorchio and G. Manara, "A Robust Differential Amplitude Codification for Chipless RFID," IEEE Microw. Wirel. Compon. Lett., vol. 25, pp. 832–834, 2015.
9. D. Pozar, "Flat Lens Antenna Concept Using Aperture Coupled Microstrip Patches," Electron. Lett., vol. 32, pp. 2109–2111, 1996.
10. M. Euler and V. Fusco, "Frequency Selective Surface Using Nested Split Ring Slot Elements as a Lens with Mechanically Reconfigurable Beam Steering Capability," IEEE Trans. Antenn. Propag., vol. 58, no. 10, pp. 3417–3421, Oct. 2010.
11. B. Chambers and K. Ford, "Tunable Radar Absorbers Using Frequency Selective Surfaces," IEEE Eleventh Int. Conf. Antenn. Propag., vol. 2, pp. 593–597, 2001.
12. P. Yuste, J. M. Rius, J. Romeu, S. Blanch, A. Heldring and E. Ubeda, "A Microwave Invisibility Cloak: The Design, Simulation and Measurement of a Simple and Effective Frequency-Selective Surface-Based Mantle Cloak," IEEE Antenn. Propag. Mag., vol. 60, no. 4, pp. 49–59, Aug. 2018.
13. W. Tang, Y. Hao and R. Mittra, "Design of a Carpet Cloak to Conceal an Antenna Located Underneath," IEEE Trans. Antenn. Propag., vol. 60, no. 9, pp. 4444–4449, Sept. 2012.
14. M. Mahmoodi, L. Van Zant and K. M. Donnell, "An Aperture Efficiency Approach for Optimization of FSS-Based Sensor Resolution," IEEE Trans. Inst. Meas., vol. 69, no. 10, pp. 7837–7845, Oct. 2020.
15. M. Mahmoodi and K. M. Donnell, "Performance Metrics for Frequency Selective Surface-Based Sensors," IEEE Sens. Lett., vol. 1, no. 6, pp. 1–4, Dec. 2017.
16. S. Milici, J. Lorenzo, A. Lazaro, R. Villarino and D. Girbau, "Wireless Breathing Sensor Based on Wearable Modulated Frequency Selective Surface," IEEE Sens. J., vol. 17, no. 5, pp. 1285–1292, Mar. 2017.
17. M. Saikia, S. Ghosh and K. V. Srivastava, "Design and Analysis of Ultrathin Polarization Rotating Frequency Selective Surface Using V-Shaped Slots," IEEE Antenn. Wirel. Propag. Lett., vol. 16, pp. 2022–2025, 2017.

18. L. Li, Y. Li, Z. Wu, F. Huo, Y. Zhang and C. Zhao, "Novel Polarization Reconfigurable Converter Based on Multilayer Frequency-Selective Surfaces," Proc. IEEE., vol. 103, no. 7, pp. 1057–1070, Jul. 2015.
19. M. Fartookzadeh and S. H. Mohseni Armaki, "Dual-Band Reflection-Type Circular Polarizers Based on Anisotropic Impedance Surfaces," IEEE Trans. Antenn. Propag., vol. 64, no. 2, pp. 826–830, Feb. 2016.
20. M. Euler, V. Fusco, R. Cahill and R. Dickie, "325 GHz Single Layer SubMillimeter Wave FSS Based Split Slot Ring Linear to Circular Polarization Convertor," IEEE Trans. Antenn. Propag., vol. 58, no. 7, pp. 2457–2459, Jul. 2010.
21. M. Hosseini and S. V. Hum, "A Semianalytical Approach to Designing High-Transparency Low-Profile Circular Polarizers," IEEE Trans. Antenn. Propag., vol. 66, no. 12, pp. 7138–7147, Dec. 2018.
22. E. Arnieri, F. Greco and G. Amendola, "A Broadband, Wide-Angle Scanning, Linear-to-Circular Polarization Converter Based on Standard Jerusalem Cross Frequency Selective Surfaces," IEEE Trans. Antenn. Propag., vol. 69, no. 1, pp. 578–583, Jan. 2021.
23. L. M.-Lopez, J. R.-Cuevas, J. I. M.-Lopez and A. E. Martynyuk, "A Multilayer Circular Polarizer Based on Bisected Split-Ring Frequency Selective Surfaces," IEEE Antenn. Wirel. Propag. Lett., vol. 13, pp. 153–156, Jan. 2014.
24. S. H. Abadi, K. Ghaemi and N. Behdad, "Ultra-Wideband, True-Time Delay Reflectarray Antennas Using Ground-Plane-Backed, Miniaturized Element Frequency Selective Surfaces," IEEE Trans. Antenn. Propag., vol. 63, no. 2, pp. 534–542, Feb. 2015.
25. S. Celozzi, R. Araneo and G. Lovat, "Electromagnetic Shielding," 1st ed. New York, NY, USA: Wiley, 2008.
26. D. Li, T. W. Li, E. P. Li and Y. J. Zhang, "A 2.5-D Angularly Stable Frequency Selective Surface Using Via-Based Structure for 5G EMI Shielding," IEEE Trans. Electromagn. Compat., vol. 60, pp. 768–775, 2018.
27. A. Tennant, W. Hurley and T. Dias, "Experimental Knitted, Textile Frequency Selective Surfaces," Electron. Lett., vol. 48, pp. 1386–1388, 2012.
28. C. N. Chiu and K. P. Chang, "A Novel Miniaturized-Element Frequency Selective Surface Having a Stable Resonance," IEEE Antenn. Wirel. Propag. Lett., vol. 8, pp. 1175–1177, 2009.
29. L. Martinez-Lopez, J. Rodriguez-Cuevas, J. I. Martinez-Lopez, and A. E. Martynyuk, "A Multilayer Circular Polarizer Based on Bisected Split-Ring Frequency Selective Surfaces," Electron. Lett., vol. 13, pp. 153–156, 2014.
30. A. H. Abdelrahman, A. Z. Elsherbeni and F. Yang, "Transmission Phase Limit of Multilayer Frequency-Selective Surfaces for Transmitarray Designs," IEEE Trans. Antenn. Propag., vol. 62, no. 2, pp. 690–697, Feb. 2014.
31. S. M. A. Momeni Hasan Abadi, M. Li and N. Behdad, "Harmonic-Suppressed Miniaturized-Element Frequency Selective Surfaces With Higher Order Band-pass Responses," IEEE Trans. Antenn. Propag., vol. 62, no. 5, pp. 2562–2571, May 2014.
32. S. M. A. M. H. Abadi and N. Behdad, "Design of Wideband, FSS-Based Multi-Beam Antennas Using the Effective Medium Approach," IEEE Trans. Antenn. Propag., vol. 62, no. 11, pp. 5557–5564, Nov. 2014.
33. N. Xu, J. Gao, J. Zhao and X. Feng, "A Novel Wideband, Low-Profile and Second Order Miniaturized Band-Pass Frequency Selective Surfaces," AIP Adv., vol. 5, 2015.

34. A. L. P. Campos, E. E. C. de Oliveira and P. H. da Fonseca Silva, "Design of Miniaturized Frequency Selective Surfaces Using Minkowski Island Fractal," J. Microw. Optoelectron. Electromagn. Appl., vol. 9, no. 19, pp. 43–49, 2010.
35. Y. Shi, W. Tang, W. Zhuang and C. Wang, "Miniaturised Frequency Selective Surface Based on 2.5-Dimensional Closed Loop," Electron. Lett., vol. 50, pp. 1656–1658, 2014.
36. Y. Yu, C. Chiu, Y. Chiou and T. Wu, "A Novel 2.5-Dimensional Ultraminiaturized-Element Frequency Selective Surface," IEEE Trans. Antenn. Propag., vol. 62, no. 7, pp. 3657–3663, Jul. 2014.
37. B. Li and Z. Shen, "Three-Dimensional Bandpass Frequency-Selective Structures With Multiple Transmission Zeros," IEEE Trans. Microw. Tech., vol. 61, no. 10, pp. 3578–3589, Oct. 2013.
38. A. K. Rashid, B. Li and Z. Shen, "An overview of Three-Dimensional Frequencyselective Structures," IEEE Antenn. Propag. Mag., vol. 56, no. 3, pp. 43–67, Jun. 2014.
39. G. I. Kiani, K. L. Ford, K. P. Esselle, A. R. Weily, C. Panagamuwa and J. C. Batchelor, "Single-Layer Bandpass Active Frequency Selective Surface," Microw. Opt. Technol. Lett., vol. 50, pp. 2149–2151, 2008.
40. B. Li and Z. Shen, "Bandpass Frequency Selective Structure With Wideband Spurious Rejection," IEEE Antenn. Wirel. Propag. Lett., vol. 13, pp. 145–148, 2014.
41. C. Huang, C. Ji, X. Wu, J. Song and X. Luo, "Combining FSS and EBG Surfaces for High-Efficiency Transmission and Low-Scattering Properties," IEEE Trans. Antenn. Propag., vol. 66, no. 3, pp. 1628–1632, Mar. 2018.
42. L. Sun, H. Cheng, Y. Zhou and J. Wang, "Broadband Metamaterial Absorber Based on Coupling Resistive Frequency Selective Surface," Opt. Express., vol. 20, pp. 4675–4680, 2012.
43. R. Anwar, L. Mao and H. Ning, "Frequency Selective Surfaces: A Review," Appl. Sci., vol. 8, no. 9, p. 1689, Sept. 2018.
44. F. Bilotti, L. Nucci and L. Vegni, "An SRR Based Microwave Absorber," Microw. Opt. Technol. Lett., vol. 48, no. 11, pp. 2171–2175, Nov. 2006.
45. N. I. Landy, S. Sajuyigbe, J. J. Mock, D. R. Smith and W. J. Padilla, "Perfect Metamaterial Absorber," Phys. Rev. Lett., vol. 100, p. 207402, May 2008.
46. H. Li, L. H. Yuan, B. Zhou, X. P. Shen, Q. Cheng and T. J. Cu, "Ultrathin Multiband Gigahertz Metamaterial Absorbers," J. Appl. Phys., vol. 110, p. 014909, Jul. 2011.
47. A. K. Singh, M. P. Abegaonkar and S. K. Koul, "Dual- and Triple-Band Polarization Insensitive Ultrathin Conformal Metamaterial Absorbers With Wide Angular Stability," IEEE Trans. Electromagn. Compat., vol. 61, no. 3, pp. 878–886, Jun. 2019.
48. J. Wang, R. Yang, J. Tian, X. Chen and W. Zhang, "A Dual-Band Absorber with Wide-Angle and Polarization Insensitivity," IEEE Antenn. Wirel. Propag. Lett., vol. 17, no. 7, pp. 1242–1246, Jul. 2018.
49. P. Jain, A. K. Singh, J. K. Pandey, S. Garg, S. Bansal, M. Agarwal, S. Kumar, N. Sardana, N. Gupta and A. K. Singh, "Ultra-Thin Metamaterial Perfect Absorbers for Single-/Dual-/Multi-Band Microwave Applications," IET Microw. Antenn. Propag., vol. 14, no. 5, pp. 390–396, Apr. 2020.
50. O. Luukkonen, F. Costa, C. R. Simovski, A. Monorchio and S. A. Tretyakov, "A thin Electromagnetic Absorber for Wide Incidence Angles and Both Polarizations," IEEE Trans. Antenn. Propag., vol. 57, no. 10, pp. 3119–3125, Oct. 2009.

51. S. Bhattacharyya and K. V. Srivastava, "An Ultra Thin Electric Field Driven LC Resonator Structure as Metamaterial Absorber for Dual Band Applications," In Proceedings URSI International Symposium on Electromagnetic Theory (EMTS), pp. 722–725, May 20–24, 2013.
52. M.-H. Li, H.-L. Yang and X.-W. Hou, "Perfect Metamaterial Absorber with Dual Band," Prog. Electromag. Res., vol. 108, pp. 37–49, Sept. 2010.
53. P. V. Tuong, J. W. Park, J. Y. Rhee, K. W. Kim, W. H. Jang, H. Cheong and Y. P. Lee, "Polarization-Insensitive and Polarization-Controlled Dual-Band Absorption in Metamaterials," J. Appl. Phys., vol. 102, p. 081122, Mar. 2013.
54. H.-X. Xu, G.-M. Wang, M.-Q. Qi, J.-G. Liang, J.-Q. Gong and Z.-M. Xu, "Triple- Band Polarization-Insensitive Wide-Angle Ultra-Miniature Metamaterial Transmission Line Absorber," Phys. Rev. B, vol. 86, p. 205104, Nov. 2012.
55. B. Bian, S. Liu, S. Wang, X. Kong, H. Zhang, B. Ma and H. Yang, "Novel Triple-Band Polarization-Insensitive Wide-Angle Ultra-Thin Microwave Metamaterial Absorber," J. Appl. Phys., vol. 114, p. 194511, Nov. 2013.
56. D. Chaurasiya, S. Ghosh, S. Bhattacharyya and K. V. Srivastava, "An Ultrathin Quad-Band Polarization-Insensitive Wide-Angle Metamaterial Absorber," Microw. Opt. Technol. Lett., vol. 57, no. 3, pp. 697–702, Mar. 2015.
57. M. Agarwal, A. K. Behera and M. K. Meshram, "Wide-Angle Quad-Band Polarisation-Insensitive Metamaterial Absorber," Electron. Lett., vol. 52, no. 5, pp. 340–342, Jan. 2016.
58. A. Bhattacharya, S. Bhattacharyya, S. Ghosh, D. Chaurasiya and K. V. Srivastava, "An Ultrathin Penta-Band Polarization-Insensitive Compact Metamaterial Absorber for Airborne Radar Applications," Microw. Opt. Technol. Lett., vol. 57, no. 11, pp. 2519–2524, Nov. 2015.
59. S. Bhattacharyya, S. Ghosh and K. V. Srivastava, "Bandwidth-Enhanced Metamaterial Absorber Using Electric Field-Driven LC Resonator for Airborne Radar Applications," Microwave Opt. Technol Lett., vol. 55, no. 9, p. 21312137, Sept. 2013.
60. J. Lee and S. Lim, "Bandwidth-Enhanced and Polarisation-Insensitive Metamaterial Absorber Using Double Resonance," Electron. Lett., vol. 47, no. 1, pp. 8–9, Jan. 2011.
61. D. Wen, H. Yang, Q. Ye, M. Li, L. Guo and J. Zhang, "Broadband Metamaterial Absorber Based on a Multi-layer Structure," Phys. Scr., vol. 88, p. 015402, Jun. 2013.
62. Y. Z. Cheng, Y. Wang, Y. Nie, R. Zhou Gong, X. Xiong and X. Wang, "Design, Fabrication and Measurement of a Broadband Polarization-Insensitive Metamaterial Absorber Based on Lumped Elements," J. Appl. Phys., vol. 111, no. 4, p. 044902, 2012.
63. S. Ghosh, S. Bhattacharyya and K. V. Srivastava, "Design, Characterisation and Fabrication of a Broadband Polarisation-Insensitive Multi-layer Circuit Ana-Logue Absorber," Microw. Antenn. Propag., vol. 10, no. 8, pp. 850–855, May 2016.
64. N. Hakla, S. Ghosh and K. V. Srivastava, "Design of Low-Profile Broadband Capacitive Circuit Absorber," Electron. Lett., vol. 52, no. 22, pp. 1825–1826, Oct. 2016.
65. B. Zhang, C. Jin and Z. Shen, "Low-Profile Broadband Absorber Based on Multimode Resistor-Embedded Metallic Strips," IEEE Trans. Microw. Theory Techn., vol. 68, no. 3, pp. 835–843, Mar. 2020.

66. W. S. Arceneaux, R. D. Akins and W. B. May, "Absorptive/Transmissive Radome," US. Patent, 5,400,043, May 21, 1995.
67. B. Yi, P. Liu, C. Yang and Y. Dong, "Analysis of Absorptive and Transmissive Radome," IEEE 6th International Symposium on Microwave, Antenna, Propagation and EMC Technologies (MAPE), pp. 616–619, 2015.
68. Y. Han, Y. Chang and W. Che, "Frequency-Selective Rasorbers: A View of Frequency-Selective Rasorbers and Their Application in Reducing the Radar Cross Sections of Antennas," IEEE Microw. Mag., vol. 23, no. 2, pp. 86–98, Feb. 2022.
69. A. A. Omar, H. Huang and Z. Shen, "Absorptive Frequency-Selective Reflection/Transmission Structures: A Review and Future Perspectives," IEEE Antenn. Propag. Mag., vol. 62, no. 4, pp. 62–74, Aug. 2020.
70. P. Mei, X. Q. Lin, J. W. Yu and P. C. Zhang, "A Band-Notched Absorber Designed With High Notch-Band-Edge Selectivity," IEEE Trans. Antenn. Propag., vol. 65, no. 7, pp. 3560–3567, Jul. 2017.
71. Y. Han, L. Zhu, Y. Chang and B. Li, "Dual-Polarized Bandpass and Band-Notched Frequency-Selective Absorbers Under Multimode Resonance," IEEE Trans. Antenn. Propag., vol. 66, no. 12, pp. 7449–7454, Dec. 2018.
72. A. Sharma, S. Ghosh and K. V. Srivastava, "A Polarization-Insensitive Band-Notched Absorber for Radar Cross Section Reduction," IEEE Antenn. Wirel. Propag. Lett., vol. 20, no. 2, pp. 259–263, Feb. 2021.
73. H. Huang and Z. Shen, "Low-RCS Reflectarray With Phase Controllable Absorptive Frequency-Selective Reflector," IEEE Trans. Antenn. Propag., vol. 67, no. 1, pp. 190–198, Jan. 2019.
74. B. Zhang, C. Jin and Z. Shen, "Absorptive Frequency-Selective Reflector Based on Bent Metallic Strip Embedded With Chip-Resistor," IEEE Trans. Antenn. Propag., vol. 68, no. 7, pp. 5736–5741, Jul. 2020.
75. B. Zhang, C. Jin, Q. Lv, J. Chen and Y. Tang, "Low-RCS and Wideband Reflectarray Antenna With High Radiation Efficiency," IEEE Trans. Antenn. Propag., vol. 69, no. 7, pp. 4212–4216, Jul. 2021.
76. F. Costa and A. Monorchio, "A Frequency Selective Radome with Wideband Absorbing Properties," IEEE Trans. Antenn. Propag., vol. 60, no. 6, pp. 2740–2747, Jun. 2012.
77. W. Yu et al., "Dual-Polarized Band-Absorptive Frequency Selective Rasorber Using Meander-Line and Lumped Resistors," IEEE Trans. Antenn. Propag., vol. 67, no. 2, pp. 1318–1322, Feb. 2019.
78. G. Q. Luo, W. Yu, Y. Yu, H. Jin, K. Fan and F. Zhu, "Broadband Dual Polarized Band-Absorptive Frequency-Selective Rasorber Using Absorptive Transmission/Reflection Surface," IEEE Trans. Antenn. Propag., vol. 68, no. 12, pp. 7969–7977, Dec. 2020.
79. Y. Yu, Z. Shen, T. Deng and G. Luo, "3-D Frequency-Selective Rasorber with Wide Upper Absorption Band," IEEE Trans. Antenn. Propag., vol. 65, no. 8, pp. 4363–4367, Aug. 2017.
80. A. Motevasselian and B. L. G. Jonsson, "Design of a Wideband Rasorber with a Polarisation-Sensitive Transparent Window," IET Microw. Antenn. Propag., vol. 6, no. 7, pp. 747–755, 2012.
81. Q. Chen, J. Bai, L. Chen and Y. Fu, "A miniaturized Absorptive Frequency Selective Surface," IEEE Antenn. Wirel. Propag. Lett., vol. 14, pp. 80–83, 2014.

82. Q. Chen, L. Liu, L. Chen, J. Bai and Y. Fu, "Absorptive Frequency Selective Surface Using Parallel LC Resonance," Electron. Lett., vol. 52, no. 6, pp. 418–419, 2016.
83. Z. Wang et al., "A High-Transmittance Frequency-Selective Rasorber Based on Dipole Arrays," IEEE Access, vol. 6, pp. 31367–31374, 2018.
84. Q. Chen, D. Sang, M. Guo and Y. Fu, "Miniaturized Frequency-Selective Rasorber with a Wide Transmission Band Using Circular Spiral Resonator," IEEE Trans. Antenn. Propag., vol. 67, no. 2, pp. 1045–1052, Feb. 2019.
85. M. Guo, Z. Sun, D. Sang, X. Jia and Y. Fu, "Design of Frequency-Selective Rasorbers Based on Centrosymmetric Bended-Strip Resonator," IEEE Access, vol. 7, pp. 24964–24970, 2019.
86. Z. Wang, J. Fu, Q. Zeng, M. Song and T. A. Denidni, "Wideband Transmissive Frequency-Selective Absorber," IEEE Antenn. Wirel. Propag. Lett., vol. 18, no. 7, pp. 1443–1447, Jul. 2019.
87. Q. Chen, M. Guo, Z. Sun, D. Sang and Y. Fu, "Polarization-Independent Frequency-Selective Rasorber with a Broadened Absorption Band," Int. J. Electron. Commun., vol. 96, pp. 178–183, 2018.
88. Y. Han and L. Xu, "Dual-Polarized Frequency-Selective Absorber with Wide Transmission Band," Microw. Opt. Technol. Lett., vol. 62, no. 3, pp. 1270–1274, 2020.
89. Q. Chen, L. Chen, J. Bai and Y. Fu, "Design of Absorptive Frequency Selective Surface with Good Transmission at High Frequency," Electron. Lett., vol. 51, no. 12, pp. 885–886, 2015.
90. K. Zhang, W. Jiang and S. Gong, "Design Bandpass Frequency Selective Surface Absorber Using LC Resonators," IEEE Antenn. Wirel. Propag. Lett., vol. 16, pp. 2586–2589, 2017.
91. M. Qu, S. Sun, L. Deng and S. Li, "Design of a Frequency-Selective Rasorber Based on Notch Structure," IEEE Access, vol. 7, pp. 3704–3711, 2019.
92. J. Xia et al., "Design of a Wideband Absorption Frequency Selective Rasorber Based on Double Lossy Layers," IEEE Trans. Antenn. Propag., vol. 68, no. 7, pp. 5718–5723, Jul. 2020.
93. H. Ye, W. Dai, X. Chen, H. Zhang, S. Bie and J. Jiang, "High-Selectivity Frequency-Selective Rasorber Based on Low-Profile Bandpass Filter," IEEE Antenn. Wirel. Propag. Lett., vol. 20, no. 2, pp. 150–154, Feb. 2021.
94. M. Guo, T. Guo, Q. Cheng, Y. Zheng and Y. Fu, "Frequency Selective Rasorber With Anisotropic Transmission Band," IEEE Antenn. Wirel. Propag. Lett., vol. 20, no. 2, pp. 155–159, Feb. 2021.
95. S. C. Bakshi, D. Mitra and F. L. Teixeira, "Wide-Angle Broadband Rasorber for Switchable and Conformal Application," IEEE Trans. Microw. Theory Techn., vol. 69, no. 2, pp. 1205–1216, Feb. 2021.
96. H. Ye, J. Wei, L. Lin, F. Liu, L. Miao, S. Bie and J. Jiang, "A Frequency Selective Surface Rasorber Based on Four Functional Layers," IEEE Trans. Antenn. Propag., vol. 69, no. 5, pp. 2768–2778, May 2021.
97. Q. Guo, J. Su, Z. Li, J. Song and Y. Guan, "Miniaturized-Element Frequency-selective Rasorber Design Using Characteristic Modes Analysis," IEEE Trans. Antenn. Propag., vol. 68, no. 9, pp. 6683–6694, Sept. 2020.
98. Q. Chen, S. Yang, J. Bai and Y. Fu, "Design of Absorptive/Transmissive Frequency-Selective Surface Based on Parallel Resonance," IEEE Trans. Antenn. Propag., vol. 65, no. 9, pp. 4897–4902, Sept. 2017.

99. A. Sharma, M. Saikia, S. Malik, S. Ghosh and K. V. Srivastava, "A Polarization Insensitive Broadband Rasorber with in-Band Transmission Response," Microw. Opt. Technol. Lett., vol. 62, no. 12, pp. 3668–3676, 2020.
100. Q. Guo, Z. Li, J. Su, L. Y. Yang and J. Song, "Dual-Polarization Absorptive/Transmissive Frequency Selective Surface Based on Tripole Elements," IEEE Antenn. Wirel. Propag. Lett., vol. 18, no. 5, pp. 961–965, May 2019.
101. A. Parameswaran, D. Kundu and H. S. Sonalikar, "A Dual-Polarized Wideband Frequency-Selective Rasorber With Low in-Band Insertion Loss and High Oblique Incidence Stability," IEEE Trans. Electromagn. Compat., vol. 63, no. 6, pp. 1820–1828, Dec. 2021.
102. Q. Yu et al., "Miniaturized Wide-Angle Rasorber With a Wide Interabsorption High Transparent Bandpass Based on Multiple 2.5-D Resonators," In IEEE Antenn. Wirel. Propag. Lett., vol. 21, no. 2, pp. 416–420, Feb. 2022.
103. M. Guo et al., "Broadband Absorptive Frequency-Selective Rasorber Based on Multilayer Resistive Sheets Using Multilayer Resonator," IEEE Trans. Antenn. Propag., vol. 70, no. 3, pp. 2009–2022, Mar. 2022.
104. W. Wan et al., "Chiral Absorber-Based Frequency Selective Rasorber With Identical Filtering Characteristics for Distinct Polarizations," IEEE Trans. Antenn. Propag., vol. 70, no. 5, pp. 3506–3514, May 2022.
105. Y. Shang, Z. Shen and S. Xiao, "Frequency-Selective Rasorber Based on Square-loop and Cross-Dipole Arrays," IEEE Trans. Antenn. Propag., vol. 62, no. 11, pp. 5581–5589, Nov. 2014.
106. Y. Han, L. Zhu, Y. Chang and B. Li, "Dual-Polarized Bandpass and Band-notched Frequency-Selective Absorbers Under Multimode Resonance," IEEE Trans. Antenn. Propag., vol. 66, no. 12, pp. 7449–7454, Dec. 2018.
107. X. Xiu, W. Che, W. Yang, Y. Han and Q. Xue, "A Highly Selective Rasorber Based on Second-Order Resonance," IEEE Trans. Antenn. Propag., vol. 19, no. 2, pp. 223–227, Feb. 2020.
108. M. Guo, Q. Chen, Z. Sun, D. Sang and Y. Fu, "Design of Dual-Band Frequency-Selective Rasorber," IEEE Antenn. Wirel. Propag. Lett., vol. 18, no. 5, pp. 841–845, May 2019.
109. X. Zhang, W. Wu, J. Huang, W. Zhang, Y. Ye and N. Yuan, "Dual-Polarized Frequency Selective Rasorber With Two Transmission Bands," IEEE Access, vol. 7, pp. 139795–139801, 2019.
110. Y. Jia, A. Wu, Y. Liu, W. Zhang and Z. Zhou, "Dual-Polarization FrequencySelective Rasorber With Independently Controlled Dual-Band Transmission Response," IEEE Antenn. Wirel. Propag. Lett., vol. 19, no. 5, pp. 831–835, May 2019.
111. X. Xiu, W. Che, W. Yang, Y. Han and Q. Xue, "Double-Polarized Dual Passband Absorptive Frequency-Selective Transmission Structure," IEEE Trans. Electromagn. Compat., vol. 62, no. 5, pp. 1951–1960, Oct. 2020.
112. S. Hu et al., "Backscattering Cross Section of Ultrawideband Antennas," IEEE Antenn. Wirel. Propag. Lett., vol. 6, pp. 70–73, 2007.
113. W. Jiang, Y. Liu, S. Gong and T. Hong, "Application of Bionics in Antenna Radar Cross Section Reduction," IEEE Antenn. Wirel. Propag. Lett., vol. 8, pp. 1275–1278, 2009.
114. Y. B. Thakare et al., "Design of Fractal Patch Antenna for Size and Radar Crosssection Reduction," IET Microw. Antenn. Propag., vol. 4, no. 2, pp. 175–181, 2010.

115. W. Wang, S. Gong, X. Wang, Y. Guan and W. Jiang, "Differential Evolution Algorithm and Method of Moments for the Design of Low-RCS antenna," IEEE Antenn. Wirel. Propag. Lett., vol. 9, pp. 295–298, 2010.
116. Y. Liu and X. Zhao, "Perfect Absorber Metamaterial for Deigning low RCS Patch Antenna," IEEE Antenn. Wirel. Propag. Lett., vol. 13, pp. 1473–1476, 2014.
117. Y. Q. Li, H. Zhang, Y. Q. Fu and N. C. Yuan, "RCS Reduction of Ridged Waveguide Slot Antenna Array Using EBG Radar Absorbing Material," IEEE Antenn. Wirel. Propag. Lett., vol. 7, pp. 473–476, 2008.
118. T. Liu, X. Cao, J. Gao, Q. Zheng, W. Li and H. Yang, "RCS Reduction of Waveguide Slot Antenna with Metamaterial Absorber," IEEE Trans. Antenn. Propag., vol. 61, no. 3, pp. 1479–1484, 2013.
119. D. Xie, X. Liu, H. Guo, X. Yang, C. Liu and L. Zhu, "A Wideband Absorber with a Multiresonant Gridded-Square FSS for Antenna RCS Reduction," IEEE Antenn. Wirel. Propag. Lett., vol. 16, pp. 629–632, 2017.
120. H. B. Baskey, E. Johari and M. J. Akhtar, "Metamaterial Structure Integrated with a Dielectric Absorber for Wideband Reduction of Antennas Radar Cross Section," IEEE Trans. Electromagn. Compat., vol. 59, no. 4, pp. 1060–1069, Aug. 2017.
121. Y. Li, K. Zheng, L. Yang and L. Du, "Wide-Band Radar Cross-Section Reduction for Antenna Using Frequency Selective Absorber," Electron. Lett., vol. 52, no. 21, pp. 1809–1811, Oct. 2016.
122. Y. Liu, K. Li, Y. Jia, Y. Hao, S. Gong and Y. J. Guo, "Wideband RCS Reduction of a Slot array Antenna Using Polarization Conversion Metasurfaces," IEEE Trans. Antenn. Propag., vol. 64, no. 1, pp. 326–331, Jan. 2016.
123. Y. Liu, Y. Hao, K. Li and S. Gong, "Wideband and Polarization Independent Radar Cross Section Reduction Using Holographic Metasurface," IEEE Antenn. Wirel. Propag. Lett., vol. 15, pp. 1028–1031, 2016.
124. Y. Liu, Y. Hao, K. Li and S. Gong, "Radar Cross Section Reduction of a Microstrip Antenna Based on Polarization Conversion Metamaterial," IEEE Antenn. Wirel. Propag. Lett., vol. 15, pp. 80–83, 2016.
125. Y. Zheng, J. Gao, X. Cao, Z. Yuan and H. Yang, "Wideband RCS Reduction of a Microstrip Antenna Using Artificial Magnetic Conductor Structures," IEEE Antenn. Wirel. Propag. Lett., vol. 14, pp. 1582–1585, 2015.
126. Y. Zhao, X.-Y. Cao, J. Gao and W.-Q. Li, "Broadband RCS Reduction and High Gain Waveguide Slot Antenna with Orthogonal Array of Polarisationdependent AMC," Electron. Lett., vol. 49, no. 21, pp. 1312–1313, Oct. 2013.
127. P. Mei, X. Q. Lin, J. W. Yu, P. C. Zhang and A. Boukarkar, "A Low Radar Cross Section and Low Profile Antenna Co-Designed With Absorbent Frequency Selective Radome," IEEE Trans. Antenn. Propag., vol. 66, no. 1, pp. 409–413, Jan. 2018.
128. P. Mei, X. Q. Lin, J. W. Yu, A. Boukarkar, P. C. Zhang and Z. Q. Yang, "Development of a Low Radar Cross Section Antenna With Band-Notched Absorber," IEEE Trans. Antenn. Propag., vol. 66, no. 2, pp. 582–589, Feb. 2018.
129. H. Huang, Z. Shen and A. A. Omar, "3-D Absorptive Frequency Selective Reflector for Antenna Radar Cross Section Reduction," IEEE Trans. Antenn. Propag., vol. 65, no. 11, pp. 5908–5917, Nov. 2017.

Chapter 2

Single-Layer Dual-Band Transmissive Rasorber

This chapter introduces a compact T-A-T configuration rasorber constructed on a single-layer dielectric substrate. The design supports dual-band transmission combined with a narrowband absorption feature. Due to its symmetric layout, the structure demonstrates polarization-independent behavior. The proposed rasorber integrates a resonant frequency selective surface (FSS) absorber on the front side and a dual-band bandpass FSS on the reverse side of the substrate. An equivalent circuit model is employed to interpret the electromagnetic behavior of the device. To experimentally verify the simulated performance, a prototype comprising a 21 × 21 array of unit cells is fabricated and tested.

2.1 DESIGN AND ANALYSIS OF THE RASORBER

Figure 2.1 illustrates the conceptual layout of the unit cell for the proposed rasorber. The design employs a single dielectric substrate layer, with distinct metallic patterns applied to each surface. On the front face, a split-ring metallic resonator is patterned, while the reverse face features a fully metalized layer with two concentric ring-shaped slots etched into it. The structure is fabricated on an FR-4 substrate characterized by a relative permittivity (ε_r) of 4.3, a loss tangent (tan δ) of 0.025, and copper cladding of 35 μm thickness. The key geometric parameters are optimized as follows: $p = 20$, $r = 6.6$, $w = 3.2$, $g = 1$, $r_1 = 4$, $r_2 = 8.5$, $w_1 = 0.8$, $w_2 = 0.8$, and $t = 0.8$ (all dimensions are in millimeters).

The development process of the proposed rasorber follows a systematic three-step approach. In the first step, a configuration featuring a split-ring resonator on the top surface and a continuous copper layer on the bottom is evaluated. This arrangement results in narrowband absorption at the resonant frequency, as illustrated in Figure 2.2(a). In the second step, a dual-band frequency selective surface (FSS) is designed by etching two concentric ring-shaped slots into a metallic sheet, with the opposite surface entirely cleared of metal, as depicted in Figure 2.2(b). The design parameters from both steps are fine-tuned such that the FSS from step-2 reflects strongly at

DOI: 10.1201/9781003644323-2

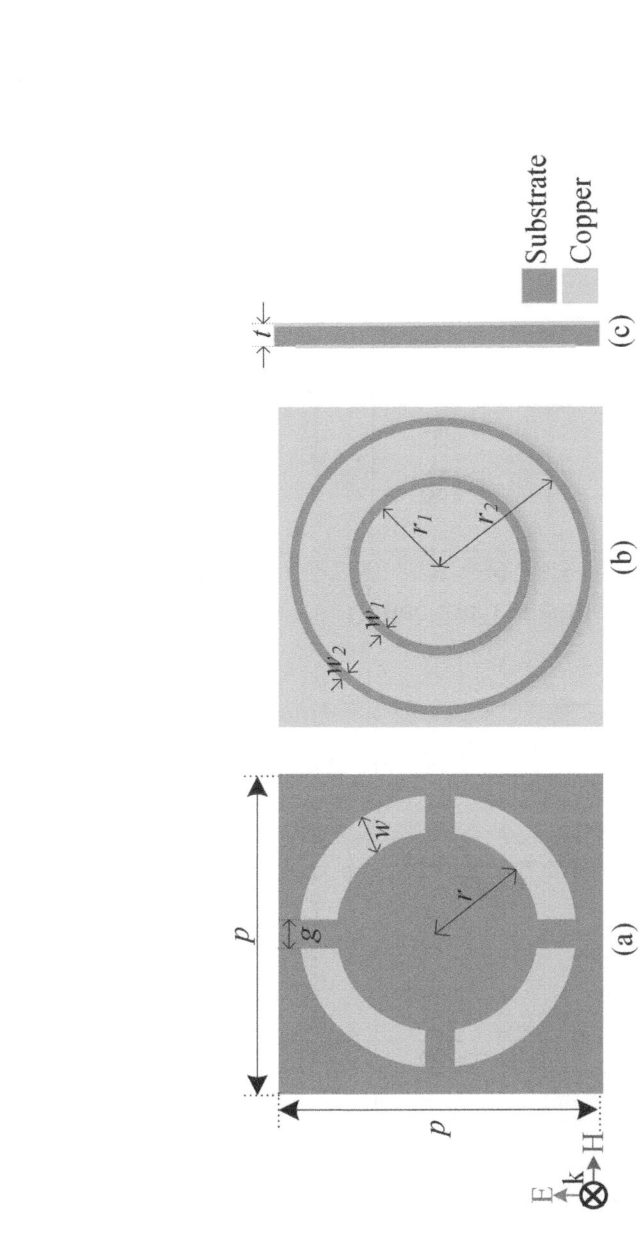

Figure 2.1 Schematic representation of the proposed FSS-based rasorber unit cell: (a) front view, (b) back view, and (c) side profile [1].

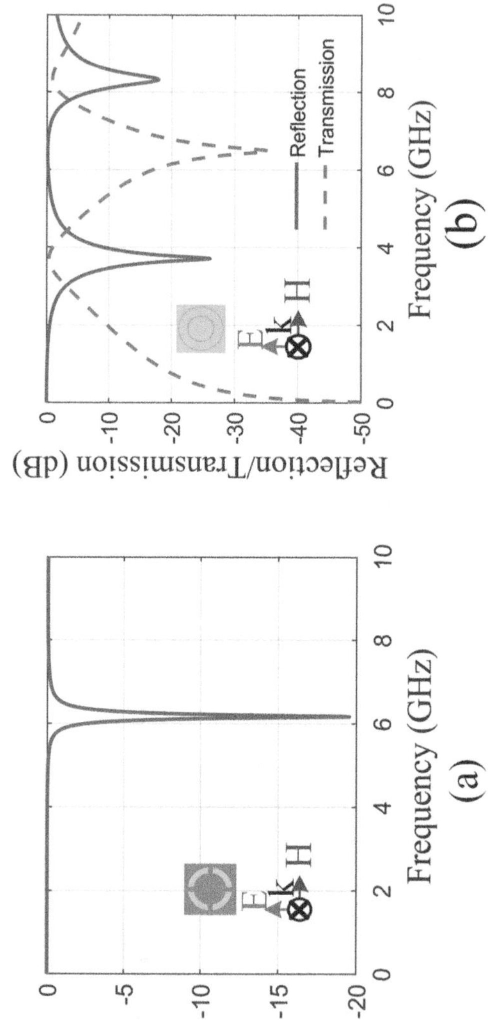

Figure 2.2 Simulated transmission and reflection responses: (a) step-1: resonant FSS absorber, (b) step-2: bandpass FSS configuration [1].

the absorption frequency identified in step-1, effectively mimicking a metallic ground plane at that frequency. Finally, in the third step, the solid copper backing in the original absorber is substituted with the dual-band FSS layer, completing the rasorber structure.

Figure 2.3 presents the equivalent circuit model (ECM) developed for the proposed rasorber. This model is constructed by integrating the individual LC representations of the metallic patterns on both the front and rear surfaces of the structure. The inductive and capacitive effects of the split-ring resonator on the front side are represented by the series combination of $L_F - C_F$. Meanwhile, the dielectric substrate separating the two layers is modeled as a lossy transmission line, with its length set equal to the physical substrate thickness and characterized by an impedance Z_d, calculated as:

$$Z_d = \frac{Z_0}{\sqrt{\epsilon_r}} \tag{2.1}$$

Here, Z_0 denotes the intrinsic impedance of free space. The term R_d accounts for the dielectric losses associated with the substrate material [2]. The propagation constant β is defined as $\beta = \beta_0\sqrt{\mu_r \epsilon_r}$, where μ_r and ϵ_r are the substrate's relative permeability and permittivity, respectively. The frequency selective behavior of the back-side dual-band FSS is modeled through a cascade of two parallel LC branches: $L_{P1} - C_{P1}$ and $L_{P2} - C_{P2}$, each corresponding to a distinct resonant frequency.

The complete ABCD matrix of the equivalent circuit model (ECM) is derived through the successive multiplication of the individual transmission matrices, expressed as:

$$\begin{bmatrix} A & B \\ C & D \end{bmatrix} = \begin{bmatrix} 1 & 0 \\ 1/Z_F & 1 \end{bmatrix} \times \begin{bmatrix} 1 & R_d \\ 0 & 1 \end{bmatrix} \times \begin{bmatrix} \cos\beta t & jZ_d\sin\beta t \\ j\sin\beta t / Z_d & \cos\beta t \end{bmatrix} \times \begin{bmatrix} 1 & 0 \\ 1/Z_P & 1 \end{bmatrix} \tag{2.2}$$

Here, Z_F and Z_P denote the impedances of the front side resonant element and the rear dual-bandpass FSS, respectively. These impedance values are computed using the following expressions:

$$Z_F = j\left(\omega L_F - \frac{1}{\omega C_F}\right) \tag{2.3}$$

$$Z_P = j\left(\frac{\omega L_{P1}}{1-\omega^2 L_{P1}C_{P1}} + \frac{\omega L_{P2}}{1-\omega^2 L_{P2}C_{P2}}\right) \tag{2.4}$$

The scattering parameters can be derived from the ABCD matrix parameters using the following transformation formula:

$$S_{11} = \frac{A + B/Z_0 - CZ_0 - D}{A + B/Z_0 + CZ_0 + D} \tag{2.5}$$

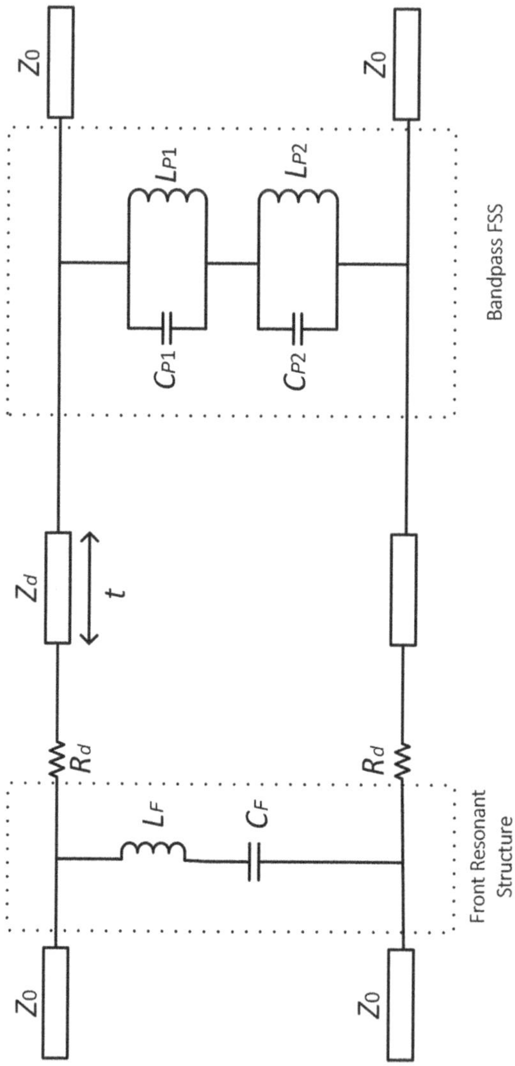

Figure 2.3 ECM of the designed FSS-based rasorber structure [1].

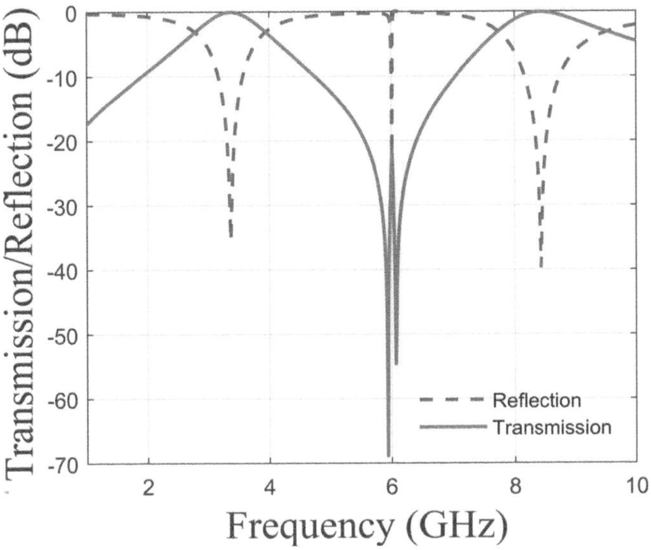

Figure 2.4 Simulated S-parameters (reflection and transmission) of the ECM [1].

$$S_{21} = \frac{2}{A + B / Z_0 + CZ_0 + D} \tag{2.6}$$

The absorptivity can be evaluated as

$$A(\omega) = 1 - |S_{11}|^2 - |S_{21}|^2 \tag{2.7}$$

The equivalent circuit model (ECM) for the proposed rasorber is simulated using Keysight ADS software. The resulting transmission and reflection characteristics are shown in Figure 2.4. The optimized component values used in the simulation are: $L_F = 51.63$ nH, $C_F = 0.0133$ pF $L_{P1} = 3.0$ nH, $C_{P1} = 0.7$ pF, $L_{P2} = 0.8$ nH, $C_{P2} = 0.437$ pF, $R_d = 4.7\ \Omega$. From the simulated results in Figure 2.4, it is evident that the $L_F - C_F$ branch, representing the front resonator, induces a resonance near 6.2 GHz, where both transmission and reflection fall below –10 dB, indicating strong absorption. Additionally, the observed transmission peaks at approximately 3.1 GHz and 8.4 GHz are attributed to the resonances created by the $L_{P1} - C_{P1}$ and $L_{P2} - C_{P2}$ branches, respectively.

The designed FSS-based rasorber was evaluated using full-wave electromagnetic simulators CST Microwave Studio and Ansys HFSS. Figure 2.5 presents the simulated S-parameters of the structure under normal incidence. The rasorber demonstrates a narrow absorption band centered around 6.2 GHz, along with two distinct transmission windows. These transmission

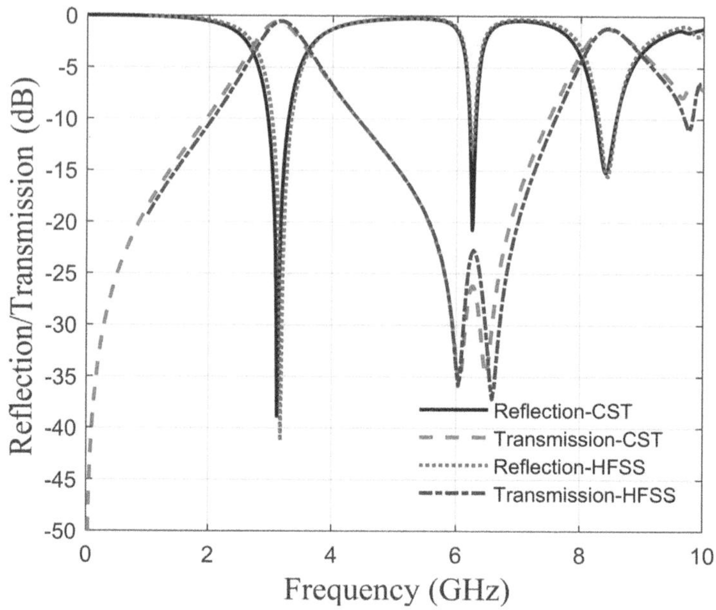

Figure 2.5 Simulated transmission and reflection responses of the designed FSS-based rasorber structure [1].

bands occur near 3.1 GHz and 8.4 GHz, with corresponding insertion losses of approximately 0.6 dB and 1.2 dB, respectively. The –10 dB bandwidths for the transmission bands span 2.95–3.3 GHz (11.20%) and 8.29–8.63 GHz (4.01%). The absorption band centered at 6.2 GHz is relatively sharp, with a –10 dB bandwidth of 6.21–6.32 GHz (1.75%), where both transmission and reflection drop below –10 dB, indicating strong absorption.

To further understand the device behavior, surface current distributions were studied at the three key frequencies, as shown in Figure 2.6. At 3.1 GHz, surface currents are predominantly concentrated along the edges of the outer ring slot, facilitating signal transmission. At 8.4 GHz, the current shifts to the inner ring slot, enabling the second transmission band. Meanwhile, at the absorption frequency of 6.2 GHz, surface currents are mainly concentrated on the edges of the front side split-ring structure, indicating localized energy dissipation and absorption.

The proposed FSS-based rasorber exhibits fourfold geometric symmetry, ensuring that its electromagnetic response remains unaffected by the polarization angle of the incident wave. This polarization-insensitive behavior is confirmed through simulated reflection and transmission characteristics across various polarization angles, as depicted in Figure 2.7. Additionally, the design is evaluated under oblique incidence scenarios, with results for different angles of incidence shown in Figure 2.8.

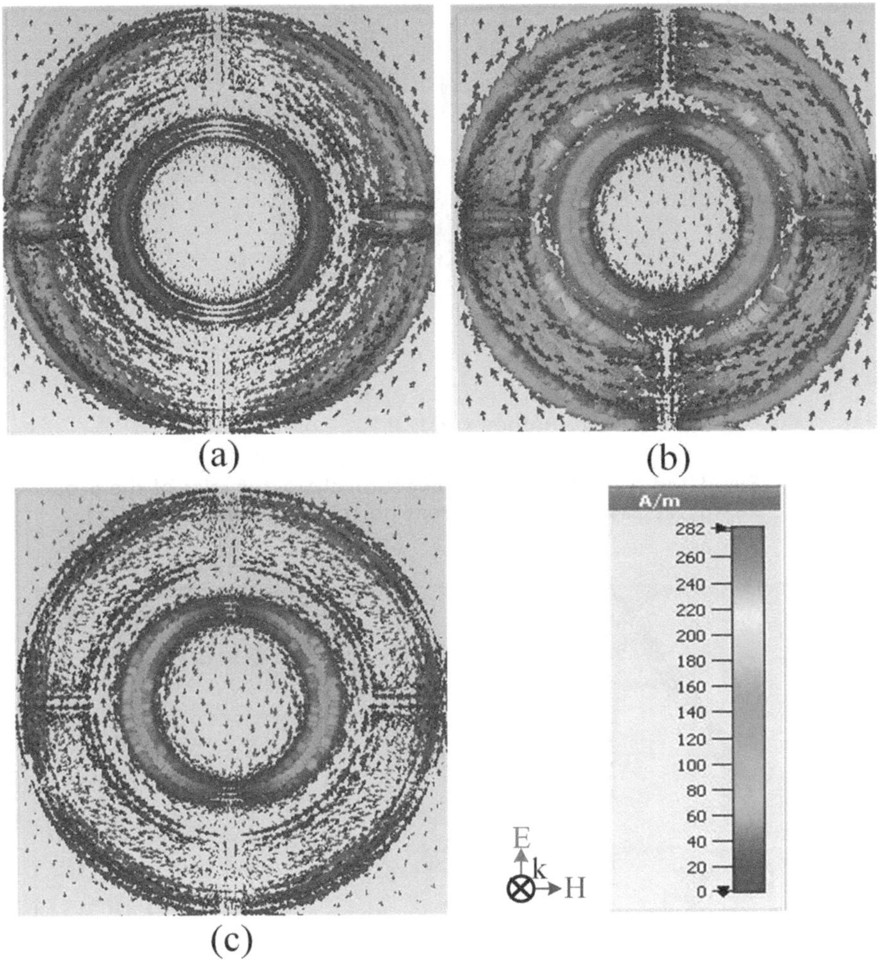

Figure 2.6 Simulated surface current patterns of the proposed rasorber at different resonant frequencies: (a) 3.1 GHz, (b) 6.2 GHz, and (c) 8.4 GHz [1].

The angular stability of the rasorber is assessed by monitoring the insertion loss at the transmission bands (centered at 3.1 GHz and 8.4 GHz) and the absorbance at the absorption band (centered at 6.2 GHz) across varying incident angles. Table 2.1 summarizes key performance metrics for incidence angles of 0°, 15°, and 30°. An insertion loss threshold of 1.5 dB is considered acceptable at the transmission frequencies. It is observed that beyond a 30° angle of incidence, particularly at 8.4 GHz, the insertion loss surpasses this threshold. Consequently, the proposed rasorber demonstrates stable and acceptable performance for incident angles up to 30°, making it suitable for practical applications with moderate angular variations.

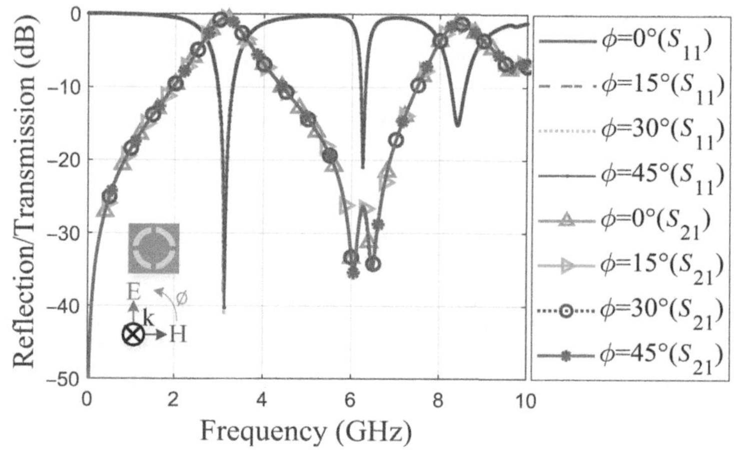

Figure 2.7 Simulated transmission and reflection characteristics of the proposed FSS-based rasorber for different polarization orientations [1].

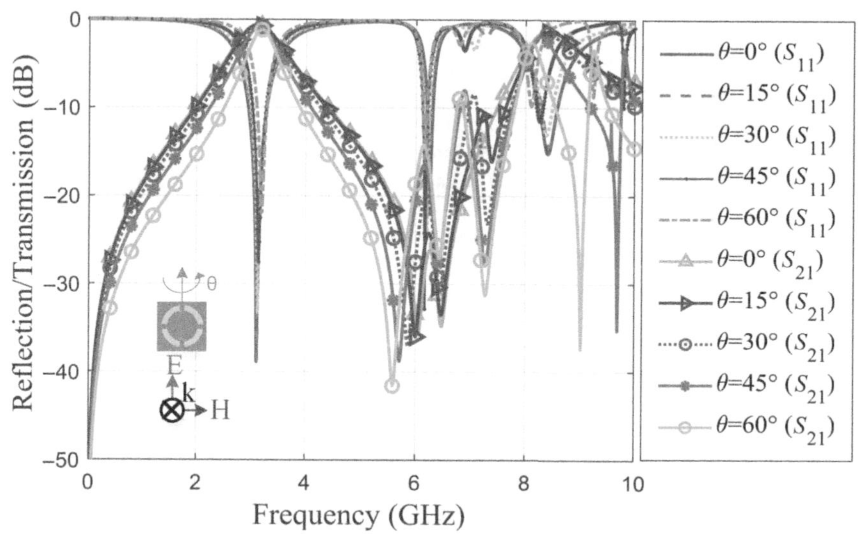

Figure 2.8 Simulated transmission and reflection behavior of the proposed FSS-based rasorber at different angles of incidence [1].

Table 2.1 Evaluated Performance Metrics under Different Incidence Angles [1]

Incidence Angle (θ)	*Insertion Loss @ 3.1 GHz*	*Absorption GHz*	*@ 6.2 Insertion Loss @ 8.4 GHz*
θ°	0.62 dB	98.91%	1.24 dB
15°	0.64 dB	99.0%	1.35 dB
30°	0.73 dB	98.63%	1.53 dB

2.2 EFFECTS OF THE DESIGN PARAMETERS ON THE PERFORMANCE OF RASORBER

The proposed rasorber features two distinct transmission windows separated by a narrow absorption band located between them. Its performance can be evaluated based on the insertion loss within the transmission bands, the absorptivity at the absorption frequency, and the relative positioning of these bands. To analyze how design parameters influence the spacing between the absorption and transmission frequencies, two transition ratios TR_1 and TR_u are introduced and defined as follows:

$$TR_l = \frac{f_{tl}}{f_a} \tag{2.8}$$

$$TR_u = \frac{f_{tu}}{f_a} \tag{2.9}$$

The parameters f_a, f_{tl}, and f_{tu} correspond to the absorption frequency, lower transmission frequency, and upper transmission frequency, respectively. Based on the surface current distribution illustrated in Figure 2.6, it is observed that f_{tl} is predominantly influenced by r_2 (the outer ring slot radius), whereas f_{tu} is primarily affected by r_1 (the inner ring slot radius). To investigate the influence of geometric variations, a parametric analysis is conducted. First, the effect of varying r_2 on transition ratio TR_1 is studied while keeping r_1 fixed. Conversely, to assess the impact on TR_u, r_1 is varied

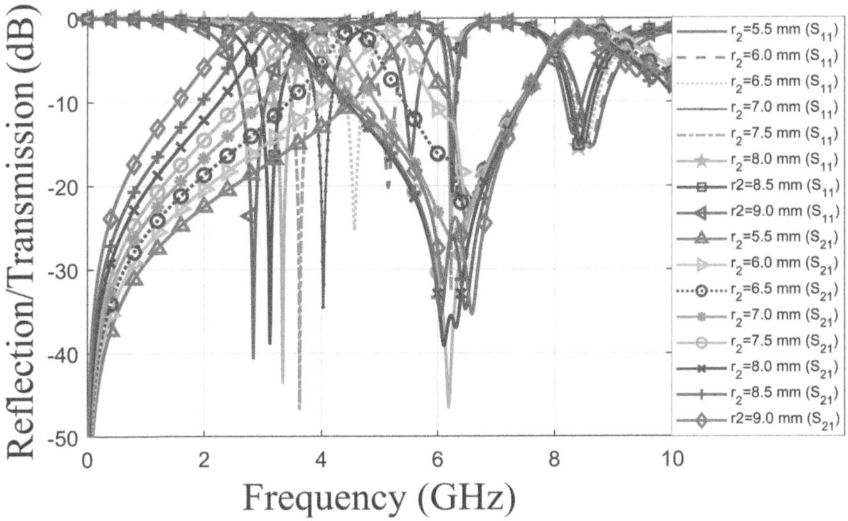

Figure 2.9 Simulated reflection and transmission characteristics for different values of r_2 at r_1 = 4.0 mm maintained constant [1].

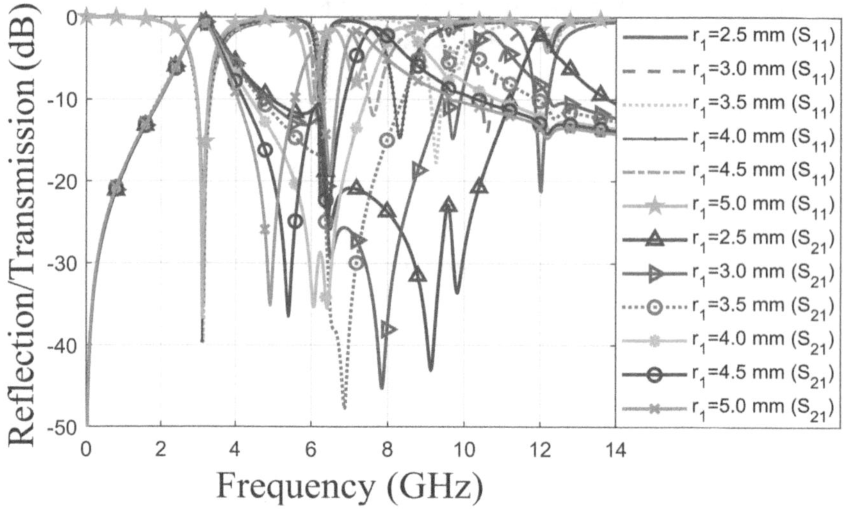

Figure 2.10 Simulated reflection and transmission characteristics for different values of r_1 at r_2 = 8.5 mm maintained constant [1].

with r_2 held constant. The corresponding sweep results for r_2 (with r_1 = 4.0 mm) are shown in Figure 2.9, and those for r_1 (with r_2= 8.5 mm) are presented in Figure 2.10. The findings from these simulations are summarized in Tables 2.2 and 2.3, respectively.

Table 2.2 indicates that reducing the outer ring slot radius r_2 from 9.0 to 5.5 mm results in a rise in the lower transmission frequency f_{tl}, shifting it from 2.8 GHz to 5.5 GHz. Similarly, Table 2.3 shows that as the inner ring slot radius r_1 decreases from 5.0 to 2.5 mm, the upper transmission frequency f_{tu} increases, reaching up to 12.0 GHz.

Based on the insights derived from Table 2.2 and Table 2.3, suitable operational ranges for both f_{tl} andf_{tu} are determined. From Table 2.2, it is evident that as r_2 decreases, the lower transmission frequency f_{t1} increases, while the upper transmission frequency f_{tu} remains largely unaffected. Notably, strong absorption at 6.2 GHz, defined here as absorptivity greater than 90%, persists until f_{tl} reaches approximately 4.5 GHz. For this analysis, acceptable performance is also characterized by maintaining the insertion loss within 2 dB across the transmission bands. Based on these conditions, the practical bounds of f_{tl} are identified as follows:

$$f_{tl(\min)} \simeq 2.8 \text{ GHz} = 0.45 f_a \tag{2.10}$$

$$f_{tl(\max)} \simeq 4.5 \text{ GHz} = 0.72 f_a \tag{2.11}$$

Table 2.2 Parametric Variation with r_2 Keeping r_1 = 4.0 mm [1]

r_1 (mm)	*I.L @ f_{tl} (GHz)*	*f_a (GHz) (Absorption %)*	*I.L @ f_{tu} (GHz)*	TR_l	TR_u
9.0	0.43 dB @ 2.8	6.2 (98.56%)	1.35 dB @ 8.4	0.45	1.35
8.5	0.61 dB @ 3.1	6.2 (98.91%)	1.24 dB @ 8.4	0.5	1.35
8.0	0.78 dB @ 3.3	6.2 (99.44%)	1.23 dB @ 8.4	0.53	1.35
7.5	0.93 dB @ 3.6	6.2 (99.92%)	1.19 dB @ 8.4	0.58	1.35
7.0	1.06 dB @ 4.0	6.2 (99.43%)	1.15 dB @ 8.4	0.64	1.35
6.5	1.19 dB @ 4.5	6.2 (95.77%)	1.22 dB @ 8.4	0.72	1.35
6.0	1.46 dB @ 5.1	6.2 (75.02%)	1.30 dB @ 8.4	0.82	1.35
5.5	2.24 dB @ 5.5	6.2 (72.42%)	1.27 dB @ 8.4	0.88	1.35

Table 2.3 Parametric Variation with r_1 Keeping r_2 = 8.5 mm [1]

r_1 (mm)	*I.L @ f_{tl} (GHz)*	*f_a (GHz) (Absorption)*	*I.L @ f_{tu} (GHz)*	TR_l	TR_u
5.0	0.65 dB @ 3.1	6.2 (48.80%)	1.54 dB @ 6.5	0.5	1.04
4.5	0.62 dB @ 3.1	6.2 (86.13%)	1.18 dB @ 7.7	0.5	1.24
4.0	0.61 dB @ 3.1	6.2 (98.91%)	1.24 dB @ 8.4	0.5	1.35
3.5	0.58 dB @ 3.1	6.2 (96.02%)	1.84 dB @ 9.2	0.5	1.46
3.0	0.58 dB @ 3.1	6.2 (93.75%)	1.79 dB @ 10.5	0.5	1.69
2.5	0.58 dB @ 3.1	6.2 (89.02%)	2.23 dB @ 12.0	0.5	1.93

Here, f_a represents the central frequency of the absorption band. Additionally, Table 2.3 shows that the lower transmission frequency f_{tl} remains nearly steady at 3.1 GHz, while the upper transmission frequency f_{tu} increases as the inner slot radius r_1 decreases. Effective absorption near 6.2 GHz is retained up to a maximum f_{tu} of approximately 10.5 GHz. Therefore, consideringf_{tl} fixed at 3.1 GHz, the valid range for f_{tu} can be defined as:

$$f_{tu(\min)} \simeq 8.4\text{ GHz} = 1.35 f_{\mathrm{a}} \tag{2.12}$$

$$f_{tu(\max)} \simeq 10.5\text{ GHz} = 1.69 f_{\mathrm{a}} \tag{2.13}$$

Therefore, with the absorption frequencyf_a fixed at 6.2 GHz, the ideal operating range for the lower and upper transmission frequencies f_{tl} and f_{tu} can be expressed as:

$$f_{tl} \simeq 0.45 f_a \text{ to } 0.72 f_a \tag{2.14}$$

$$f_{tu} \simeq 1.35 f_a \text{ to } 1.69 f_a \tag{2.15}$$

In the presented rasorber design, the transition ratio TR_l can be adjusted from 0.72 to 0.45 by varying the outer slot radius r_2 from 6.5 to 9 mm, while keeping the inner slot radius r_1 constant at 4.0 mm. Similarly, altering r_1 from 3.0 to 4.0 mm with r_2 fixed at 8.5 mm leads to a variation in TR_u from 1.69 to 1.35. This indicates that the spectral spacing between the transmission bands and the absorption band can be effectively tuned through appropriate selection of the ring slot radii r_1 and r_2.

2.3 FABRICATION AND MEASUREMENTS

To experimentally validate the simulated findings, measurements were conducted on a fabricated prototype of the proposed FSS-based rasorber. A 21 × 21 array of unit cells was implemented on a 0.8 mm thick FR-4 substrate. The fabricated structure, having an overall dimension of 460 × 460 mm, is shown in the photograph in Figure 2.11.

The experimental evaluation of the fabricated prototype was conducted at the anechoic chamber, using a free-space measurement approach. A pair of broadband horn antennas, connected to a portable Anritsu MS2028C vector network analyzer, was employed to measure the reflection (S_{11}) and transmission (S_{21}) parameters. The measurement setup is illustrated in Figure 2.12. For reflection measurements, a metallic plate was initially used as a reference to establish a baseline response. This plate was subsequently replaced with the fabricated prototype, and the resulting measurements were recorded. The effective reflection coefficient was determined by taking the difference between the reflection data of the metallic reference and that of the prototype. Similarly, for transmission measurements, the reference was obtained by recording the signal transmitted directly between the two horn antennas without any intervening object. The effective transmission coefficient was then calculated by comparing the transmission data from the direct path and the prototype. The measured data are compared with the simulated results in Figure 2.13, showing a good correspondence and validating the accuracy of the proposed design.

Additionally, measurements on the fabricated prototype were performed for different oblique angles of incidence. The resulting measured and simulated responses under these varying angles are presented in Figure 2.14, demonstrating a satisfactory correlation between experimental and simulated data.

A comparative assessment of the proposed FSS-based rasorber with earlier reported designs is provided in Table 2.4. Unlike some existing designs [3–5], the proposed rasorber offers polarization-insensitive performance and utilizes a compact, single-layer structure. Moreover, it shows improved insertion loss characteristics across both transmission bands compared to previously published rasorbers in [3–5].

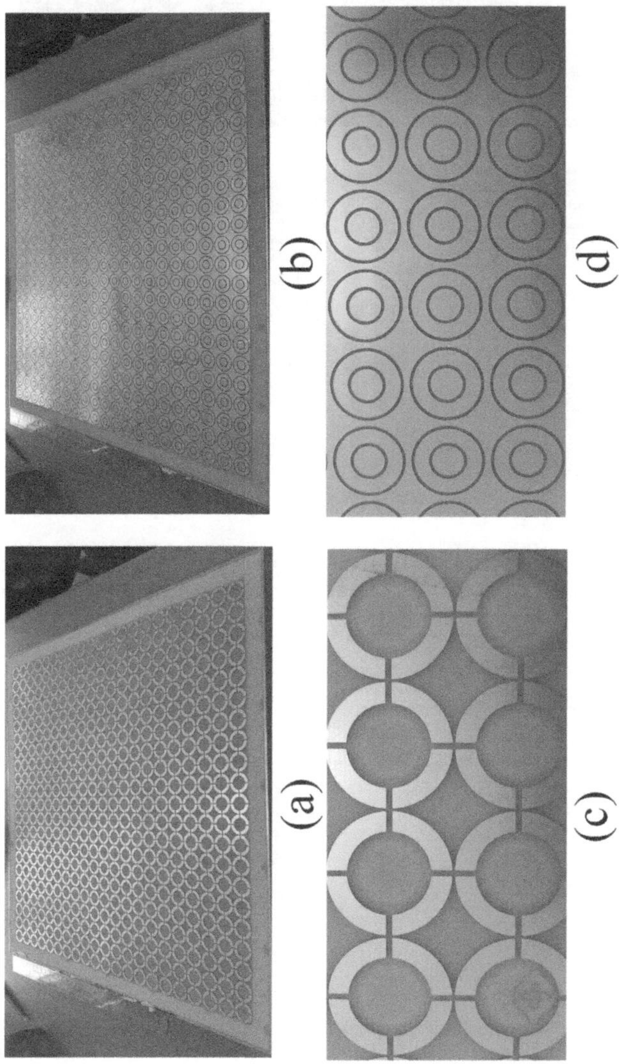

Figure 2.11 Image of the fabricated prototype: (a) top view, (b) bottom view, (c) magnified view of the front surface, and (d) magnified view of the rear surface [1].

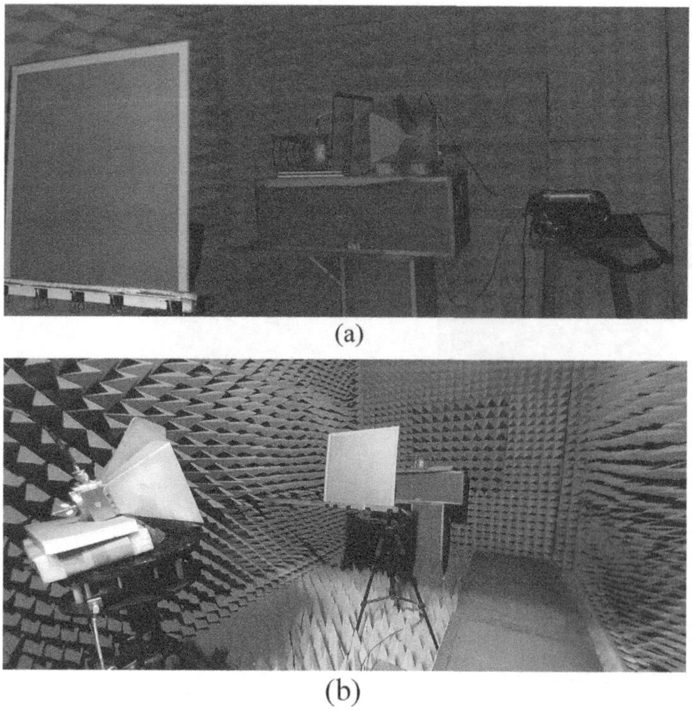

Figure 2.12 Measurement setup illustrations: (a) setup for reflection coefficient (S_{11}) measurement and (b) setup for transmission coefficient (S_{21}) measurement [1].

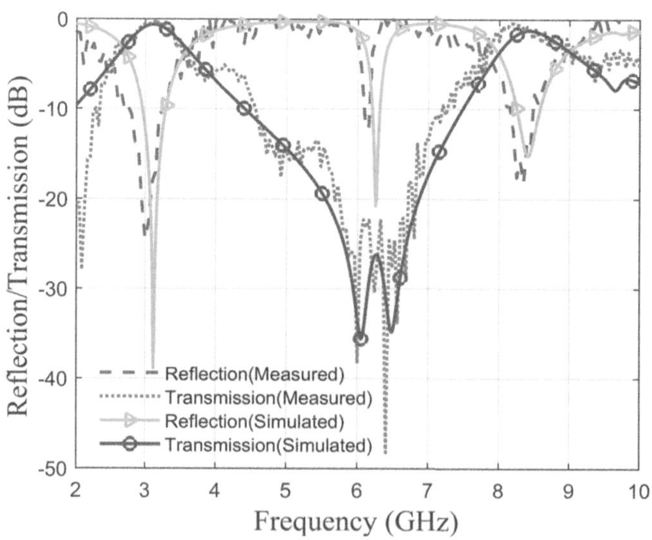

Figure 2.13 Simulated versus measured results for reflection and transmission coefficients [1].

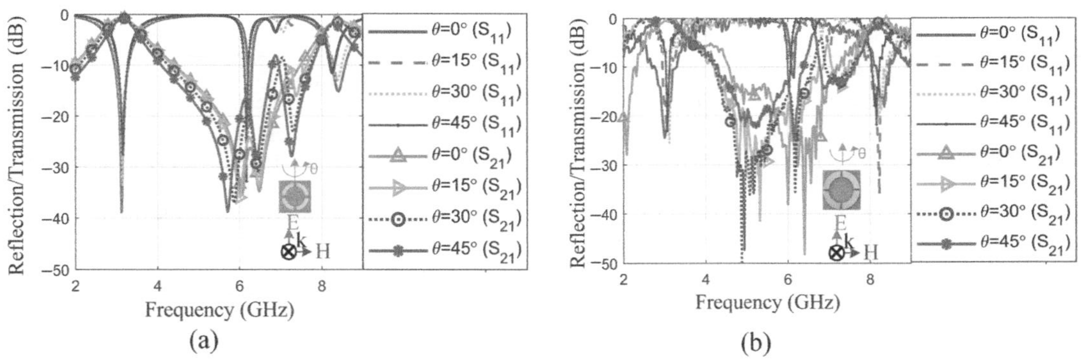

Figure 2.14 Reflection and transmission characteristics of the proposed rasorber at different angles of incidence: (a) simulation results (b) experimental measurements [1].

Table 2.4 Comparative Performance Analysis of the Proposed FSS-Based Rasorber [1]

Ref.	*Absorption*	*Transmission with I.L.*	*Thickness*	*Unit Cell Size*	*Polarization*
[3]	Broadband (4.8–6.81GHz)	I.L. < 1 dB @ < 1.5 dB	$0.04\lambda_L$(2 layers)	$0.0432\lambda_L$	Dual-polarized
[4]	Broadband (2.6–5.25, 6.05–8.5 GHz)	0.48 dB @ 5.7 GHz	$0.134\lambda_L$ (2 layers)	$0.234\lambda_L$	Dual-polarized
[5]	Broadband (4.02– 6.27, 8.21–12.02 GHz)	2.30 dB @ 7.20 GHz, 1.69 dB @ 13.05 GHz	$0.107\lambda_L$ (2 layers)	$0.2412\lambda_L \times 0.1206\lambda_L$	Polarization sensitive
This Work	Narrowband @ 6.2 GHz	0.6 dB @ 3.1 GHz, 1.2 dB @ 8.4 GHz	$0.00789\lambda_L$ (1 layer)	$0.197\lambda_L$	Polarization insensitive

2.4 CONCLUSION

This chapter presented a single-layer T-A-T-type frequency selective rasorber that achieves narrowband absorption at 6.2 GHz, flanked by two transmission windows centered at 3.1 and 8.4 GHz. The structure is developed by integrating a resonant absorber on one side and a dual-band bandpass frequency selective surface (FSS) on the opposite side of a single dielectric substrate. An equivalent circuit model (ECM) is formulated to interpret the electromagnetic behavior of the rasorber. By adjusting the design parameters of the bandpass FSS, an optimal range for the lower and upper transmission bands is established with respect to the central absorption frequency. A prototype of the structure was fabricated, and the performance was experimentally validated, confirming the effectiveness of the proposed design.

REFERENCES

1. M. M. Zargar et al., "Polarisation-Insensitive Dual-Band Transmissive Rasorber Designed on a Single Layer Substrate," IET Microw. Antenn. Propag., vol. 14, no. 11, pp. 1296–1303, 2020.
2. S. Ghosh and K. V. Srivastava, "An Equivalent Circuit Model of FSS-Based Metamaterial Absorber Using Coupled Line Theory," IEEE Antenn. Wirel. Propag. Lett., vol. 14, pp. 511–514, 2015.
3. W. Yu et al., "Dual-Polarized Band-Absorptive Frequency Selective Rasorber Using Meander-Line and Lumped Resistors," IEEE Trans. Antenn. Propag., vol. 67, no. 2, pp. 1318–1322, Feb. 2019.
4. M. Qu, S. Sun, L. Deng and S. Li, "Design of a Frequency-Selective Rasorber Based on Notch Structure," IEEE Access, vol. 7, pp. 3704–3711, 2019.
5. M. Guo, Q. Chen, Z. Sun, D. Sang and Y. Fu, "Design of Dual-Band Frequency-Selective Rasorber," IEEE Antenn. Wirel. Propag. Lett., vol. 18, no. 5, pp. 841–845, May 2019.

Chapter 3

Single-Layered Flexible Multiband Rasorbers

In the preceding chapter, a T-A-T-type rasorber was presented, featuring a single narrowband absorption between two transmission bands, implemented on a 0.8 mm thick substrate. While effective in demonstrating the T-A-T response, the design offered only single-band absorption and did not meet the physical thickness constraints required for conformal or flexible platform integration. In this chapter, advanced A-T-T-A- and A-T-A-T-A-type rasorbers are proposed, fabricated on an ultra-thin 0.25 mm flexible substrate. The A-T-T-A configuration achieves dual-band transmission and dual-band absorption (2T2A) by combining a dual-resonant frequency selective surface (FSS) absorber with a dual-bandpass FSS on opposite sides of a single dielectric layer. To enhance functionality, an additional resonant structure is incorporated, resulting in an A-T-A-T-A configuration that exhibits dual-band transmission and triple-band absorption (2T3A). These proposed rasorber designs are ultra-thin, polarization-insensitive, single-layered, and conformal, making them highly suitable for practical stealth applications.

3.1 DESIGN AND ANALYSIS OF THE RASORBER

A generalized theoretical representation of a frequency selective surface (FSS)-based rasorber is illustrated schematically in Figure 3.1. When an electromagnetic wave impinges on the rasorber, its response – whether transmission, reflection, or absorption – depends on the frequency-dependent interaction between the wave and the structure. In the transmission regime, the rasorber is designed to selectively allow wave propagation while suppressing reflections. Conversely, absorption occurs when both transmitted and reflected powers are significantly attenuated within a specific frequency band. This condition of minimal reflection and transmission corresponds to maximum absorption and can be expressed mathematically by the following relation:

$$A = 1 - R - T$$

DOI: 10.1201/9781003644323-3

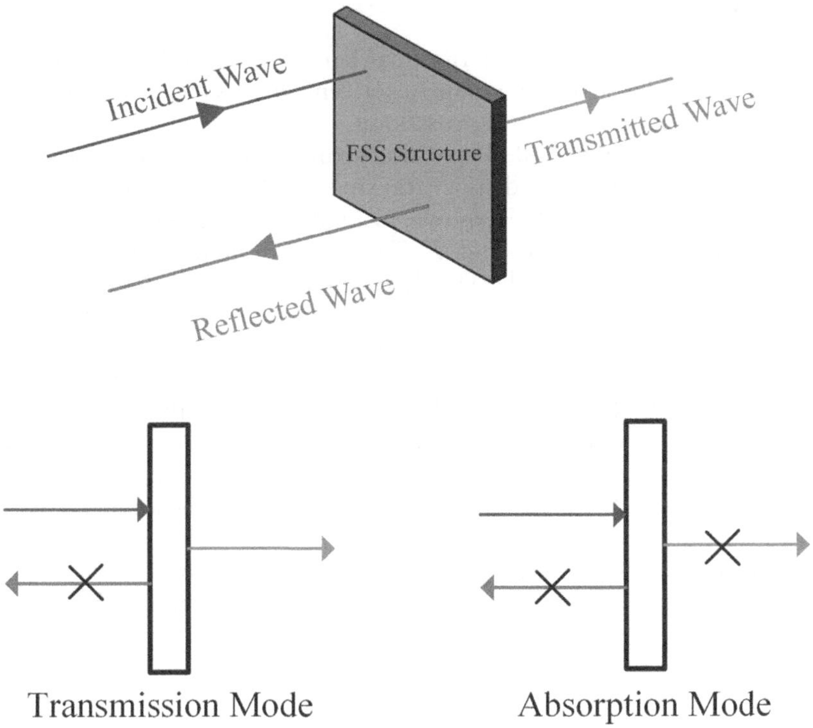

Figure 3.1 Diagrammatic illustration of FSS-based rasorber [1].

Here, A, R, and T represent the absorption, reflection, and transmission coefficients, respectively. In the configuration of a frequency selective surface (FSS)-based rasorber, the conventional metallic ground plane typically used in absorber designs is substituted with a bandpass FSS. A fundamental design criterion is that the bandpass FSS must exhibit reflective behavior at the designated absorption frequencies, thereby enabling strong resonant absorption while preserving high transmission in the desired passbands. To achieve this dual functionality, a structured three-phase design methodology is adopted. In the initial phase, a dual-band bandpass FSS is synthesized and characterized to define the spectral transmission windows. Subsequently, in the second phase, various lossy resonant geometries are investigated to generate absorption bands that are spectrally aligned with the stopbands of the bandpass FSS, ensuring minimal transmission and reflection at those frequencies. In the final phase, the absorber layer and the bandpass FSS are integrated on opposing sides of a single dielectric substrate to realize a compact, single-layer rasorber structure with the intended transmission–absorption characteristics. The proposed design is implemented on an ultra-thin, flexible Taconic dielectric substrate with a thickness of 0.254 mm, relative permittivity (ε_r), loss tangent ($\tan\delta$), and copper cladding thickness of 0.035 mm.

3.1.1 Bandpass FSS

The unit cell configuration of the dual-band bandpass FSS is illustrated in Figure 3.2. The structure comprises a central cross-loop-shaped slot of length lb_1, along with quarter cross-loop slots of length lb_2 positioned at each corner of the unit cell. These slot geometries independently support resonance, resulting in two distinct transmission bands. The full-wave simulation of the proposed FSS, shown in Figure 3.3(a), exhibits dual passbands centered at 9.7 GHz and 13.0 GHz, attributed to the central and corner loop slots, respectively. The design evolution of the bandpass FSS is presented in Figures 3.3(b) and 3.3(c), where the transmission characteristics of a simple cross-slot and a modified cross-loop slot are compared. In the basic cross-slot configuration, a single passband appears at approximately 14.3 GHz (Figure 3.3(b)). By introducing a metallic cross of width w_{b1} within the slot, forming the cross-loop geometry, the resonance shifts downward to 9.7 GHz (Figure 3.3(c)). This frequency reduction is primarily due to the increased effective slot length introduced by the looped structure. Moreover, the cross-loop configuration offers improved spectral selectivity and a sharper passband response, which is desirable for rasorber designs to ensure reflective behavior in adjacent absorption bands. The addition

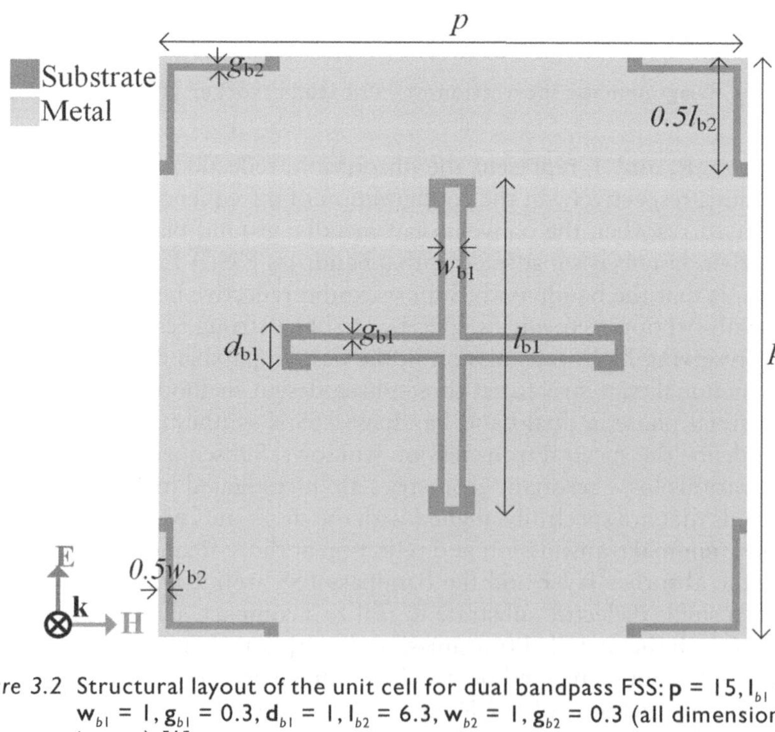

Figure 3.2 Structural layout of the unit cell for dual bandpass FSS: $p = 15$, $l_{b1} = 8.5$, $w_{b1} = 1$, $g_{b1} = 0.3$, $d_{b1} = 1$, $l_{b2} = 6.3$, $w_{b2} = 1$, $g_{b2} = 0.3$ (all dimensions are in mm) [1].

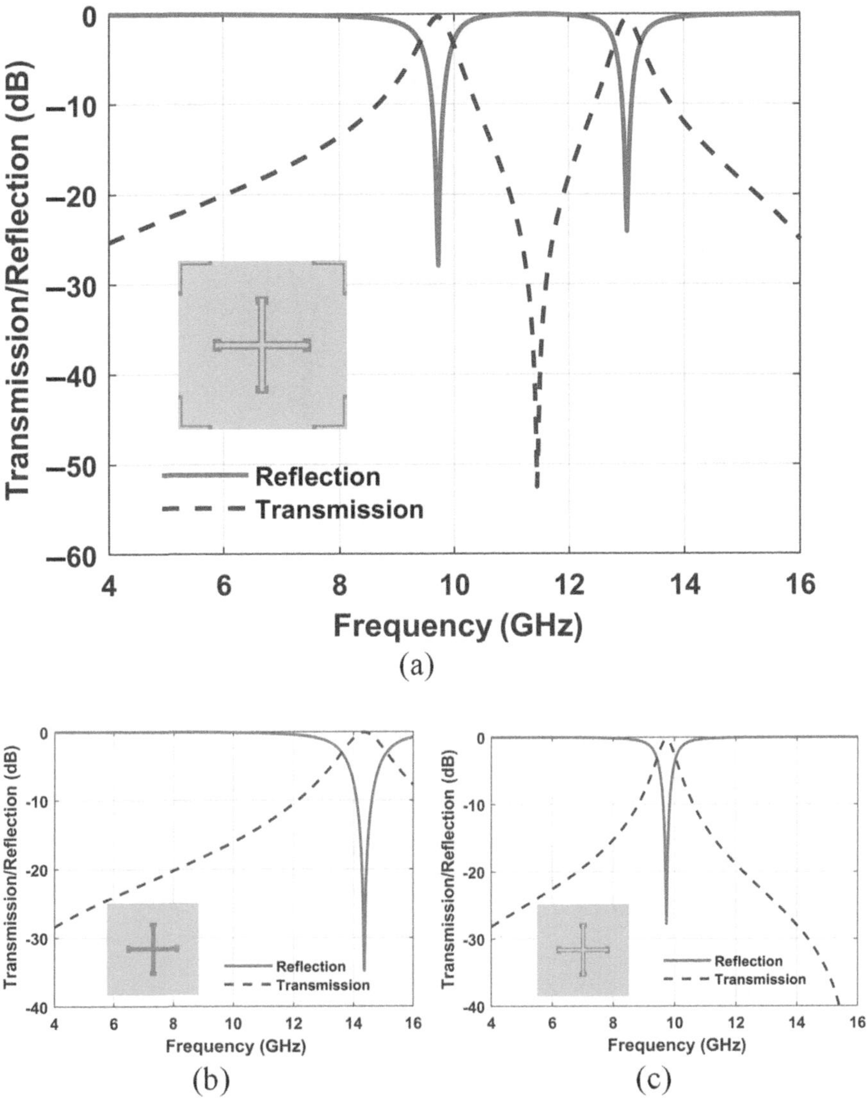

Figure 3.3 Simulated S-parameters of (a) the dual-bandpass FSS, (b) cross-slot-based bandpass FSS, and (c) cross-loop-slot-based bandpass FSS [1].

of corner cross-loop slots introduces a second passband at a higher frequency (13.0 GHz). The sensitivity of each passband to its corresponding slot dimension is analyzed in Figures 3.4(a) and 3.4(b). Varying the central cross-loop length lb_1 from 7.5 mm to 9.5 mm shifts the lower passband from 10.9 GHz to 8.9 GHz, with minimal effect on the upper passband.

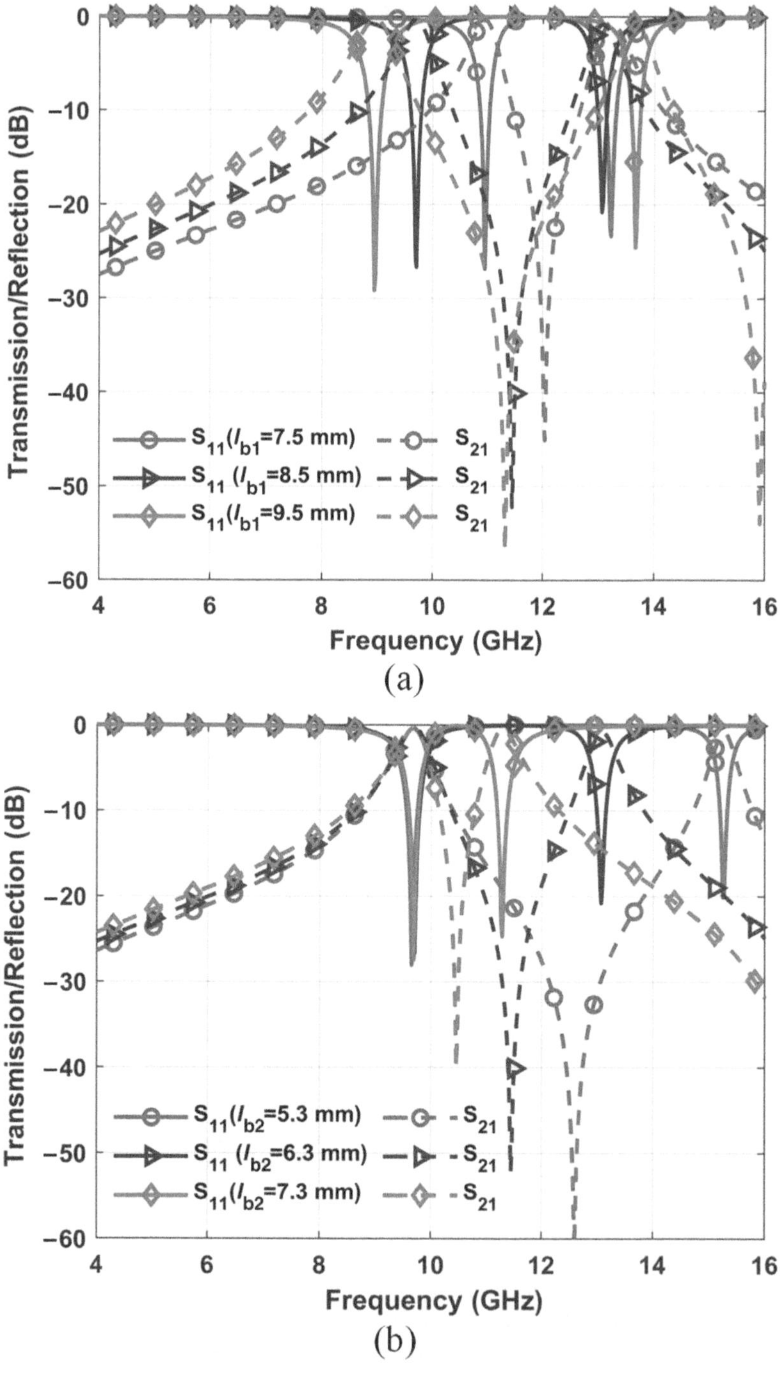

Figure 3.4 Simulation results of the dual-bandpass FSS response by varying (a) l_{b1}, and (b) l_{b2} [1].

Conversely, increasing the corner loop length lb_2 from 5.3 mm to 7.3 mm shifts the upper passband from 15.2 GHz to 11.2 GHz, while the lower passband remains largely unaffected.

3.1.2 FSS Absorber

The FSS absorber is engineered such that it exhibits strong absorption at specific frequencies where the previously designed bandpass FSS behaves as a highly reflective surface. The unit cell layout of the dual-band FSS absorber, illustrated in Figure 3.5, comprises two resonant elements – a metallic cross resonator enclosed within a metallic ring resonator patterned on the top surface of the dielectric substrate, while the bottom side is fully covered with a continuous metallic ground plane. The ring resonator is characterized by inner and outer radii r_1 and $r_1 + w_1$, respectively, whereas the cross resonator has arm length l_3 and width w_3. These two distinct resonators are designed to resonate independently at two separate frequencies, resulting in dual-band absorption. The resonant frequencies are strategically chosen such that one lies below and the other above the two transmission bands of the bandpass FSS (centered at 9.7 GHz and 13.0 GHz), ensuring spectral separation between transmission and absorption bands. The simulated reflection

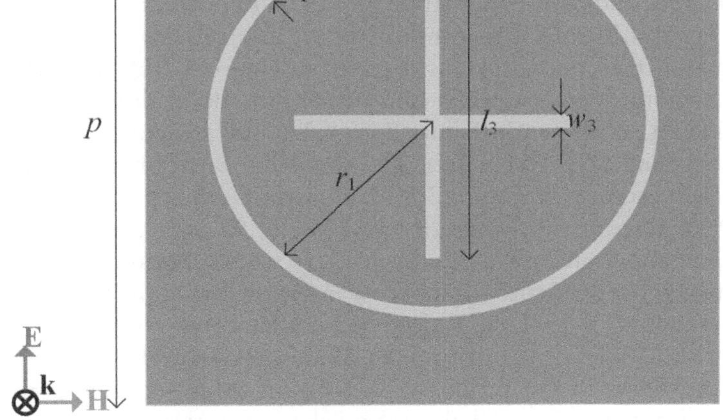

Figure 3.5 Unit cell representation of the dual-band FSS absorber: p = 15, r_1 = 5.0, w_1 = 0.5, l_3 = 7.0, w_3 = 0.4 (all dimensions are in millimeters) [1].

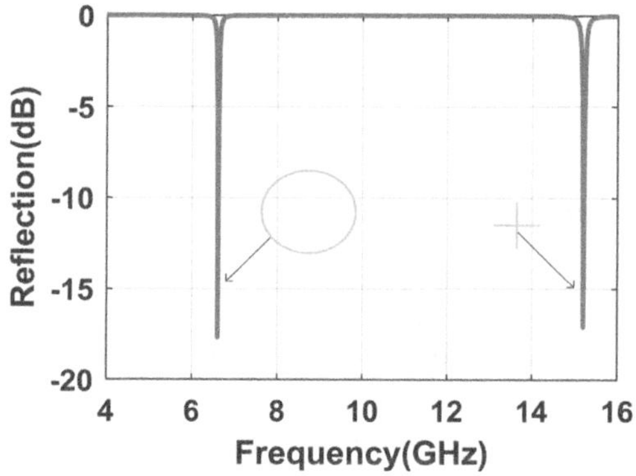

Figure 3.6 Simulated reflection profile of the dual-band FSS absorber [1].

coefficient of the absorber, presented in Figure 3.6, confirms the presence of two well-defined absorption notches. The higher-frequency absorption at 15.1 GHz is attributed to the cross resonator, while the lower-frequency absorption at 6.6 GHz arises from the ring resonator. This arrangement ensures that the absorption bands effectively bracket the transmission bands of the bandpass FSS, forming the basis for the desired A-T-T-A rasorber configuration.

3.1.3 Proposed 2T2A and 2T3A Rasorbers

The schematic of the proposed 2T2A rasorber configuration is presented in Figure 3.7. The dual-band absorption and dual-band transmission functionality are achieved by integrating the previously developed dual-band FSS absorber and dual-band bandpass FSS on opposite sides of a 0.254 mm thick Diclad substrate. The front side of the unit cell features printed metallic ring and cross resonators, as depicted in Figure 3.7(b), while the reverse side incorporates etched center and corner cross-loop slots within the copper cladding, as shown in Figure 3.7(c). The full structure is modeled and analyzed using CST EM Solver. The simulated S-parameters of the 2T2A rasorber, presented in Figure 3.8, demonstrate efficient dual-band operation. Peak absorptivity values of 93.62% and 92.29% are obtained at the lower and upper absorption frequencies, f_{al} = 6.4 GHz and f_{au} = 14.7 GHz, respectively. The structure exhibits low insertion loss at the transmission frequencies, with f_{tl} = 9.6 GHz and f_{tu} = 13.4 GHz corresponding to minimum transmission losses of 0.2 dB and 0.4 dB, respectively. These results confirm the effectiveness of the proposed rasorber in achieving high-performance dual-band transmission and absorption characteristics.

As observed in Figure 3.8, a reflective band is present between the two transmission bands centered at 9.6 GHz and 13.4 GHz, characterized by a

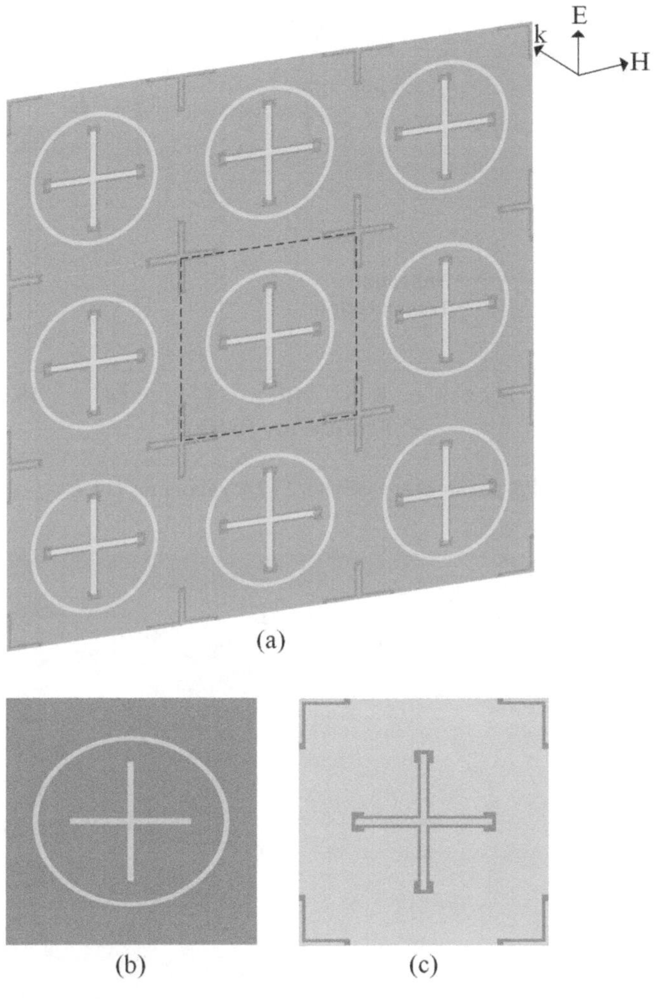

Figure 3.7 (a) 3D view of the 2T2A rasorber, (b) front face, and (c) rear face of the unit cell [1].

significant transmission dip. This intermediate reflective region provides an opportunity to introduce an additional absorption band. To exploit this, the 2T2A rasorber is modified to realize a third absorption band within this intermediate frequency range, thereby forming a structure with dual-band transmission and triple-band absorption (2T3A). The modified unit cell, depicted in Figure 3.9, incorporates an additional split-ring resonator alongside the existing ring and cross resonators. The split-ring is defined by inner and outer radii r_2 and $r_2 + w_2$, respectively, with a gap spacing denoted by g. This resonator is designed to exhibit resonance specifically within the

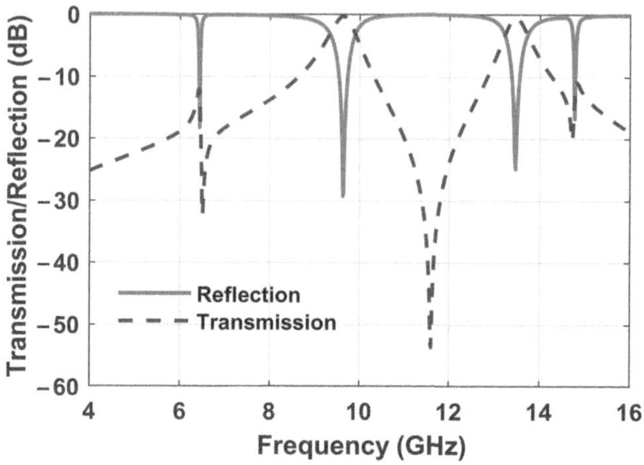

Figure 3.8 Computed S-parameters of the 2T2A rasorber [1].

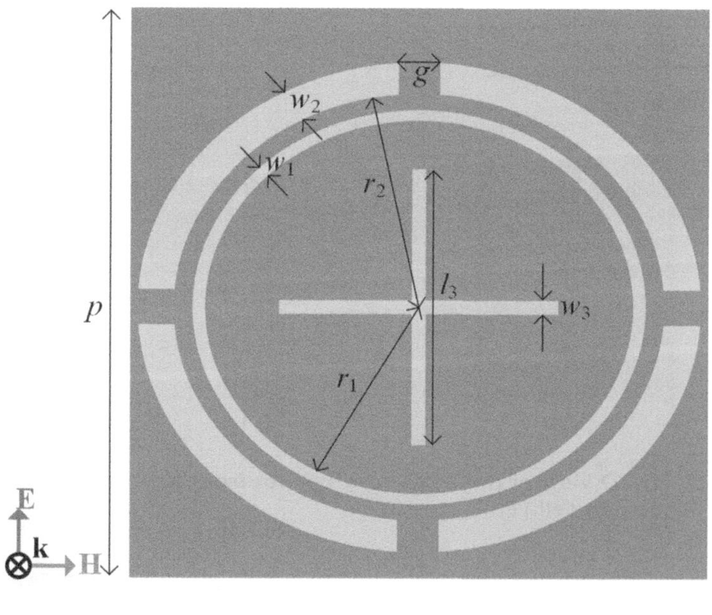

Figure 3.9 Front layout of the unit cell of the 2T3A rasorber: p = 15, r_1 = 5.0, w_1 = 0.5, l_3 = 7.0, w_3 = 0.4, r_2 = 5.9, w_2 = 0.9, w_2 = 0.9, g = 0.8 (dimensions are in mm) [1].

reflective band between the two transmission windows. A perspective view of the resulting 2T3A rasorber structure is shown in Figure 3.10. The simulated S-parameter response, presented in Figure 3.11, confirms the emergence of a third absorption band at the intermediate frequency f_{am} = 11.1 GHz, in

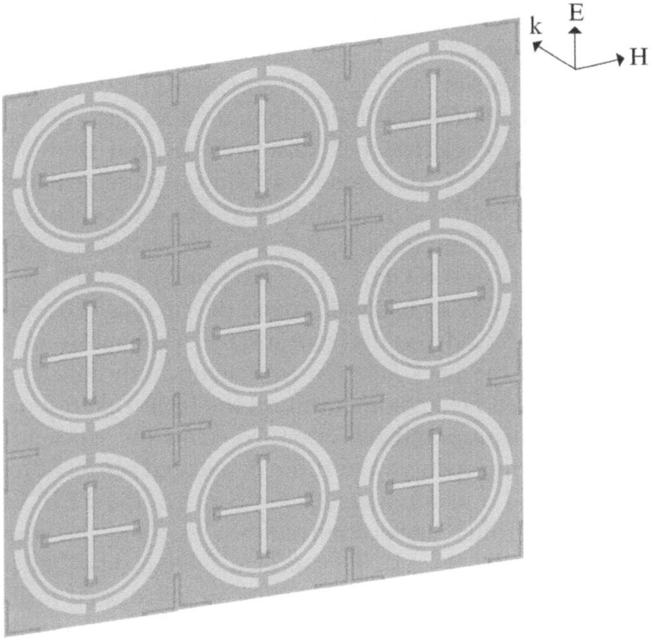

Figure 3.10 3D perspective view of the 2T3A rasorber structure [1].

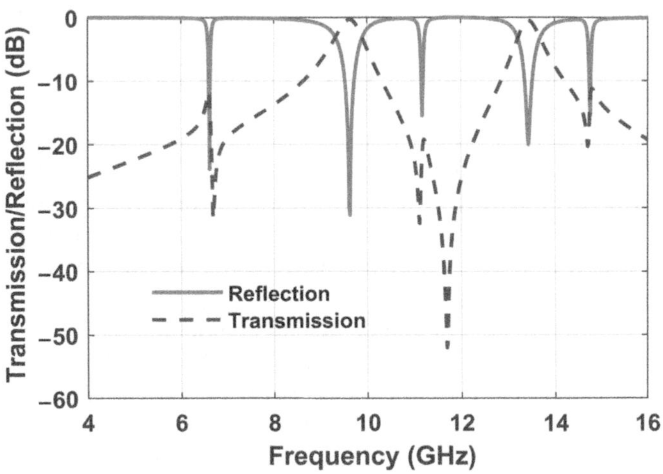

Figure 3.11 Simulated transmission and reflection behavior of the 2T3A rasorber [1].

addition to the previously obtained lower and upper absorption bands at 6.6 GHz and 14.7 GHz, respectively. The corresponding absorption peaks are 94.00%, 96.81%, and 91.91%, respectively. The two transmission bands are maintained at 9.6 GHz and 13.4 GHz, with low insertion losses of 0.3 dB and 0.5 dB, respectively. These results validate the successful extension of the original 2T2A rasorber to a high-performance 2T3A configuration.

The operational mechanism of the proposed 2T3A rasorber is elucidated through the analysis of electric field distributions and the corresponding equivalent circuit model (ECM). Figure 3.12 illustrates the scalar electric field intensity (in V/m) on both the front and back sides of the rasorber unit cell at the key frequencies of interest 6.6, 9.6, 11.1, 13.4, and 14.7 GHz. At the absorption frequencies of 6.6 GHz, 11.1 GHz, and 14.7 GHz, a strong concentration of electric fields is observed around the circular ring, split-ring, and cross resonators, respectively, on the front surface of the substrate. In contrast, the field intensity on the back side remains significantly lower at these frequencies. This localized field confinement on the front side resonators, along with minimal transmission and reflection, confirms the efficient absorption of the incident electromagnetic wave at the designated resonant frequencies.

The absorption and transmission characteristics of the 2T3A rasorber are further elucidated through surface current analysis and the corresponding equivalent circuit model (ECM). The surface current distributions at both the absorption and transmission frequencies are illustrated in Figure 3.13, providing insight into the underlying electromagnetic behavior. The ECM representations for the 2T2A and 2T3A rasorbers are depicted in Figure 3.14. In the circuit model, the front side metallic resonators are modeled using series RLC branches, while the slot-type resonators on the back side are represented by parallel LC tank circuits. The inclusion of the split-ring resonator in the 2T3A configuration is accounted for by an additional parallel LC branch, denoted as the L_3–C_3 network in Figure 3.14(b). This modular circuit representation captures the frequency selective absorption and transmission behavior of the proposed rasorber structures.

At the lower absorption frequency of 6.6 GHz, the surface current distribution is predominantly concentrated along the circular ring resonator, as shown in Figure 3.13(a). The time-varying induced currents generate an inductive effect along the conductive ring, while the spacing between adjacent elements introduces a capacitive effect. Together, these reactive components form a series LC resonance, which is modeled by the L_1–C_1 branch in the equivalent circuit shown in Figure 3.14. At the intermediate absorption frequency of 11.1 GHz, the surface currents are localized around the split-ring resonator on the front side of the structure, as illustrated in Figure 3.13(b). The presence of the split gap introduces an additional capacitive element, contributing to the overall series *LC* behavior, which is represented by the L_3–C_3 branch in the ECM. For the higher absorption frequency at 14.7 GHz, the surface current is concentrated along the metallic cross resonator situated at the center of the unit cell, as shown in Figure 3.13(c). This structure also exhibits a series *LC* response, which is captured by the L_2–C_2 branch in the ECM. These three *LC* resonant branches collectively account for the triple-band absorption behavior of the proposed 2T3A rasorber.

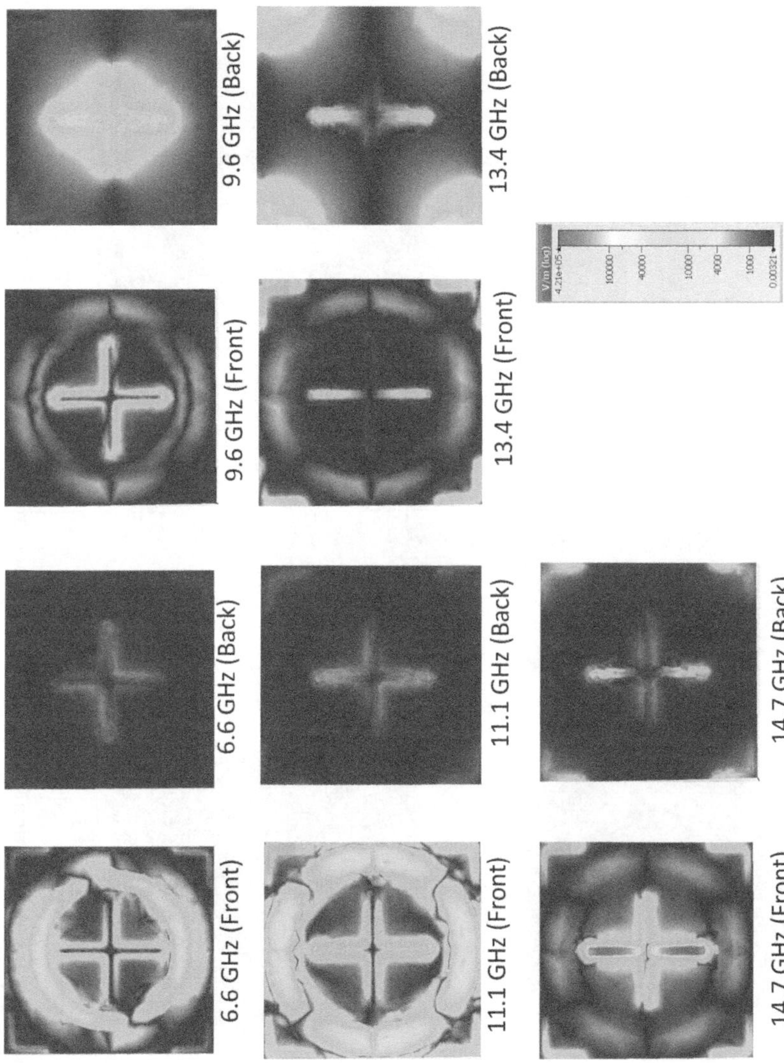

Figure 3.12 Distribution of electric field in the 2T3A rasorber [1].

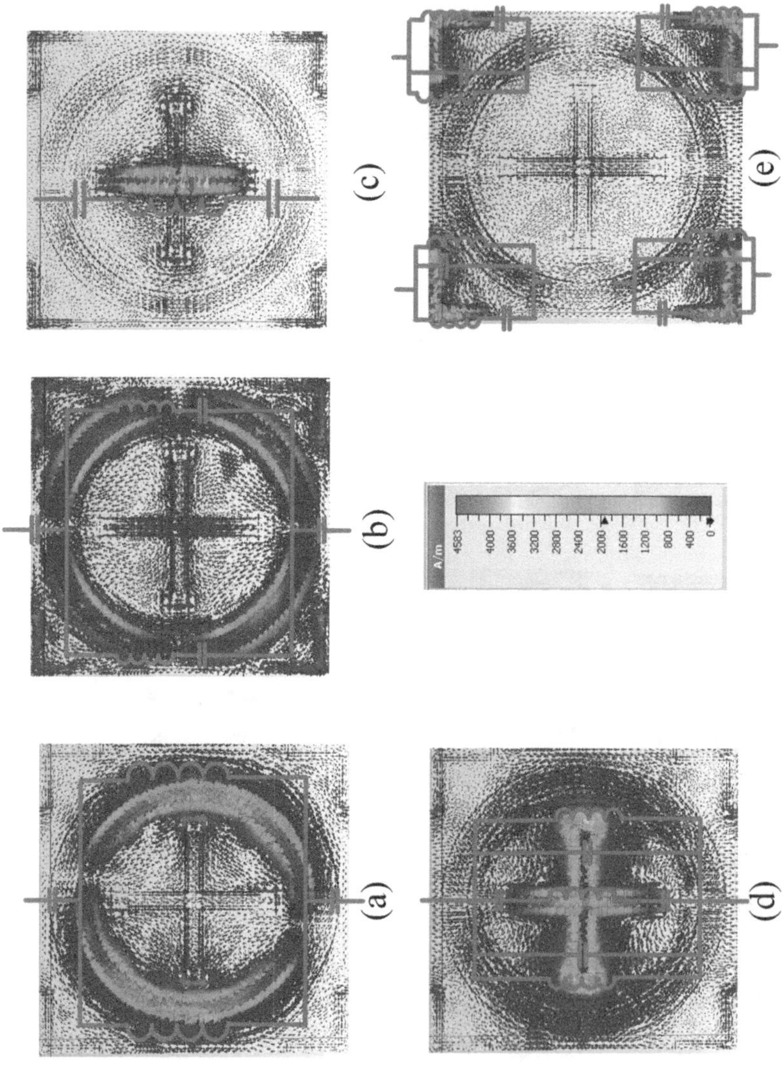

Figure 3.13 Surface current distribution of the 2T3A rasorber at (a) 6.6 GHz, (b) 11.1 GHz, (c) 14.7 GHz, (d) 9.6 GHz, and (e) 13.4 GHz [1].

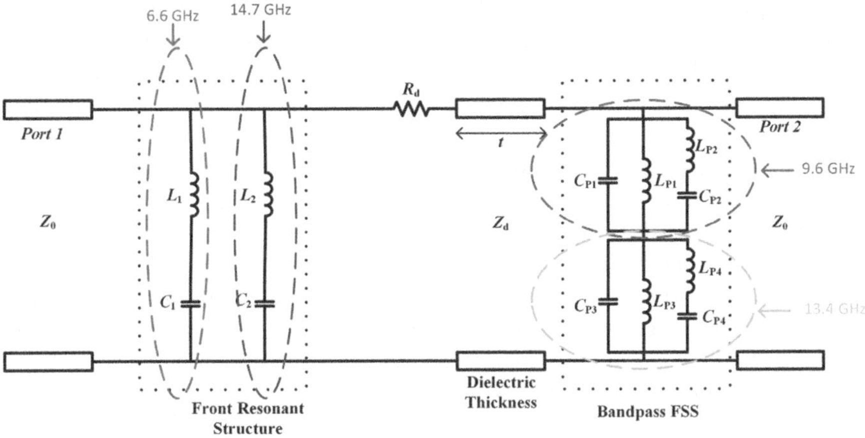

(a)

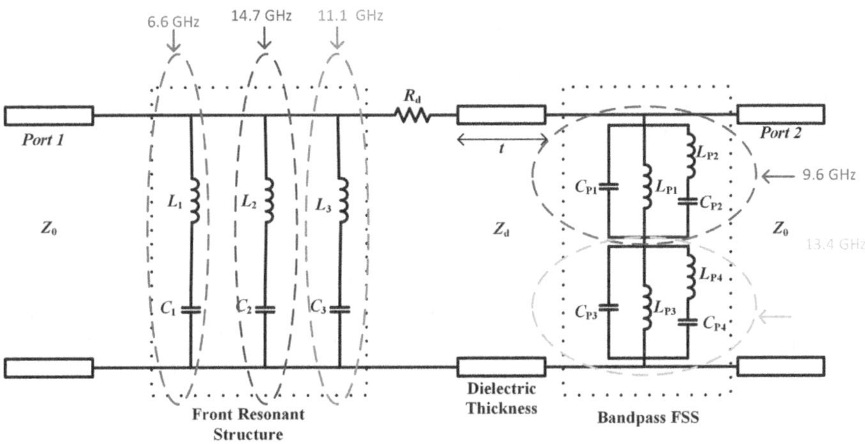

(b)

Figure 3.14 ECM of the proposed (a) 2T2A, and (b) 2T3A rasorbers ($\mathbf{L}_1$ = 91.827 nH, $\mathbf{L}_2$ = 15.260 nH, $\mathbf{L}_3$ = 25.553 nH, $\mathbf{C}_1$ = 0.006 pF, $\mathbf{C}_2$ = 0.007 pF, $\mathbf{C}_3$ = 0.007 pF, $\mathbf{R}_d$ = 1.791 Ω, $\mathbf{Z}_d$ = 254.173 Ω, $\mathbf{L}_{P1}$ = 0.145 nH, $\mathbf{C}_{P1}$ = 0.100 pF, $\mathbf{L}_{P2}$ = 0.180 nH, $\mathbf{C}_{P2}$ = 0.850 pF, $\mathbf{L}_{P3}$ = 0.120 nH, $\mathbf{C}_{P3}$ = 0.010 pF, $\mathbf{L}_{P4}$ = 0.311 nH, $\mathbf{C}_{P4}$ = 0.315 pF) [1].

At the transmission frequencies of 9.6 GHz and 13.4 GHz, the electric field intensity is predominantly concentrated along the edges of the center and corner cross-loop slots, respectively, located on the back side of the structure, as illustrated in Figure 3.13. In contrast to the absorption bands, minimal field concentration is observed on the front side metallic resonators at these frequencies. The strong localization of the electric field around the slot peripheries facilitates resonance within the slot structures, enabling efficient transmission while suppressing reflections. This field distribution confirms the frequency selective transmission behavior associated with the slot-based bandpass FSS.

At the transmission frequencies of 9.6 GHz and 13.4 GHz, the surface current distribution is primarily induced along the edges of the center and corner cross-loop slots, respectively, on the back side of the structure, aligning with the orientation of the incident electric field. The resonant behavior observed at these frequencies arises from the combined inductive and capacitive effects generated by the geometry and current flow along the slot peripheries. These electromagnetic interactions are modeled using *LC* networks, as depicted in Figures 3.13(d) and 3.13(e). Specifically, the resonances at 9.6 GHz and 13.4 GHz are represented by parallel LC branches (L_{P1}, L_{P2}, C_{P1}, C_{P2}) for the lower band and (L_{P3}, L_{P4}, C_{P3}, C_{P4}) for the upper band in the equivalent circuit model shown in Figure 3.14. These parallel resonant circuits accurately capture the frequency selective transmission characteristics of the slot-based FSS.

The dielectric substrate is modeled as a transmission line segment with a physical length equal to the substrate thickness t, and a characteristic impedance Z_d defined by:

$$Z_d = \frac{Z_0}{\sqrt{\epsilon_r}} \tag{3.1}$$

Here, Z_0 denotes the characteristic impedance of free space, and R_d represents the equivalent resistance accounting for the dielectric loss of the substrate [2]. The overall ABCD matrix of the equivalent circuit model (ECM) is computed through the cascaded multiplication of individual transmission matrices, expressed as:

$$\begin{bmatrix} A & B \\ C & D \end{bmatrix} = \begin{bmatrix} 1 & 0 \\ Y_F & 1 \end{bmatrix} \times \begin{bmatrix} 1 & R_d \\ 0 & 1 \end{bmatrix} \times \begin{bmatrix} \cos\beta t & jZ_d \sin\beta t \\ j\sin\beta t / Z_d & \cos\beta t \end{bmatrix} \times \begin{bmatrix} 1 & 0 \\ Y_P & 1 \end{bmatrix} \tag{3.2}$$

whereY_F, Y_P is the admittance of front resonant structure and dual bandpass FSS at the back side of substrate, respectively. Also, the propagation constantβ is given as $\beta = \beta_0\sqrt{\mu_r \epsilon_r}$ (μ_r and ϵ_r are the relative permeability and permittivity of the dielectric substrate, respectively).

The S-parameters of the equivalent circuit model (ECM) are derived from the corresponding ABCD parameters using the standard transformation equations:

$$S_{11} = \frac{A + B / Z_0 - CZ_0 - D}{A + B / Z_0 + CZ_0 + D} \tag{3.3}$$

$$S_{21} = \frac{2}{A + B / Z_0 + CZ_0 + D} \tag{3.4}$$

From S-parameters obtained in Equations (3.3) and (3.4), the absorptivity$A(\omega)$ can be calculated using the following formula:

$$A(\omega) = 1 - |S_{11}|^2 - |S_{21}|^2 \tag{3.5}$$

The equivalent circuit models (ECMs) of the proposed 2T2A and 2T3A rasorbers were analyzed using Keysight ADS. The simulated reflection and transmission responses are presented in Figure 3.15. For the 2T2A rasorber, the ECM results shown in Figure 3.15(a) demonstrate dual absorption resonances at 6.5 GHz and 14.8 GHz, along with two transmission bands centered at 9.5 GHz and 13.5 GHz. In the case of the 2T3A rasorber, the inclusion of an additional resonant branch in the ECM enables a third absorption band at 11.1 GHz, as illustrated in Figure 3.15(b).

The polarization response of the proposed 2T2A and 2T3A rasorbers was investigated by simulating their performance under varying polarization angles of the incident electromagnetic wave, as illustrated in Figure 3.16. For both designs, the simulated transmission and reflection coefficients exhibit negligible variation with changes in polarization angle, thereby confirming the polarization-insensitive characteristics of the proposed rasorbers.

The polarization behavior of the proposed 2T3A rasorber is examined by analyzing the cross-polarized (t_{ij}) and co-polarized (t_{ii}) transmission coefficients. Figure 3.17(a) presents the co- and cross-polarization transmission responses for both *x*- and *y*-polarized incident waves. The cross-polarized transmission components (t_{yx} and t_{xy}) are negligible in comparison to their co-polarized counterparts (t_{xx} and t_{yy}), indicating that the polarization of the transmitted wave remains identical to that of the incident wave. This confirms that the proposed transmissive rasorber preserves the polarization state of the incident electromagnetic wave. Additionally, the polarization conversion ratio (PCR) plot, shown in Figure 3.17(b), illustrates that PCR_{yy} remains close to 1, while PCR_{xy} approaches 0 at both transmission frequencies, further validating the absence of polarization conversion in the structure.

The angular stability of the proposed rasorbers is evaluated under oblique incidence for TE polarization, as illustrated in Figure 3.18. For both the 2T2A and 2T3A configurations, the reflection and transmission coefficients exhibit stable and acceptable performance up to an incident angle of 40°, maintaining effective absorption and transmission at the designated frequencies. However, beyond 40°, a noticeable degradation is observed in the absorption performance specifically at 14.7 GHz, while the responses at other operational frequencies remain largely unaffected up to 60°. Consequently, the angular stability of the proposed rasorber structures is effectively limited to an incident angle of 40°.

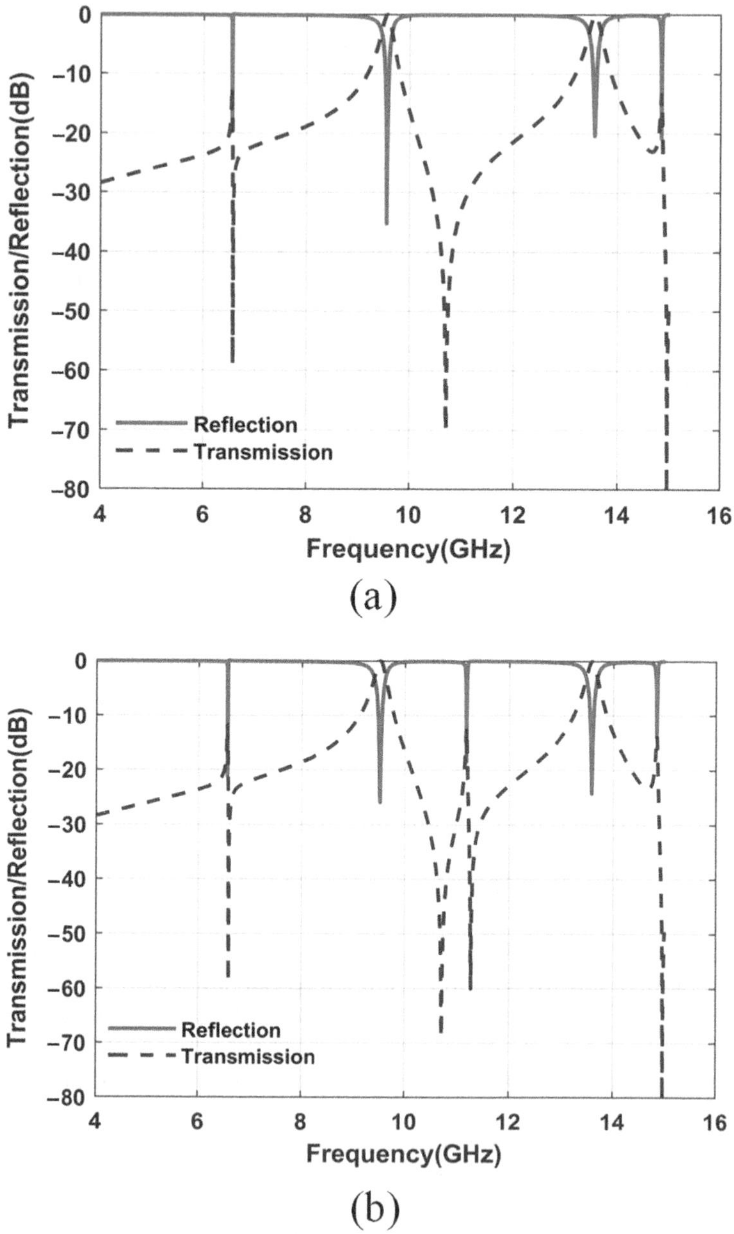

Figure 3.15 Simulated response for ECM of the proposed (a) 2T2A and (b) 2T3A rasorbers [1].

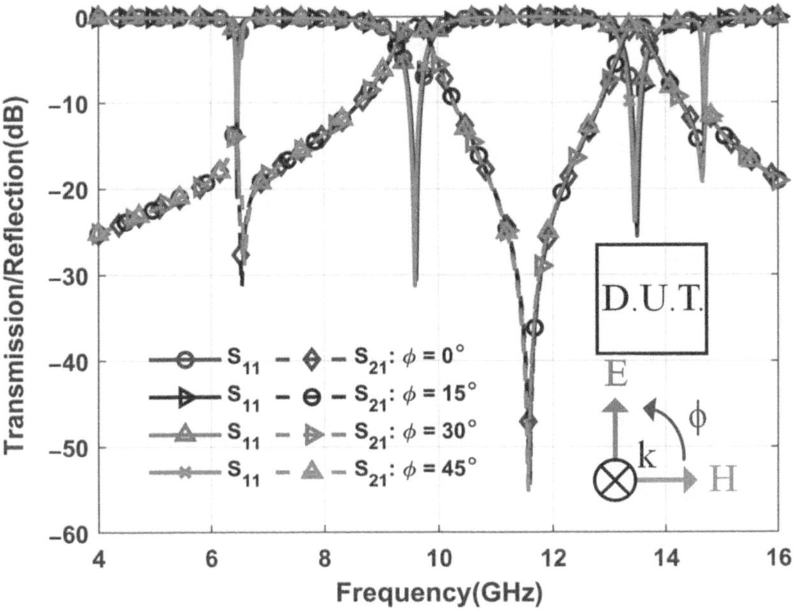

(a)

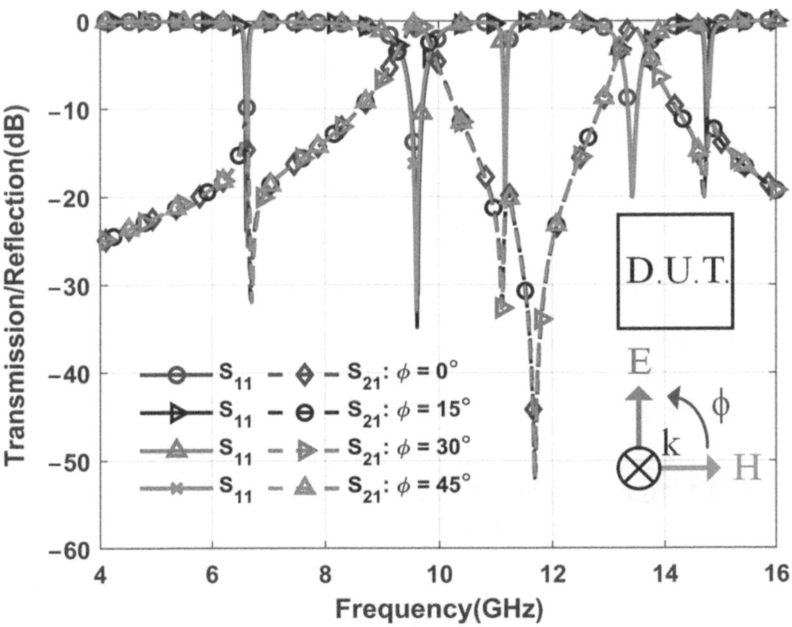

(b)

Figure 3.16 Simulated response under various polarization angles for the proposed (a) 2T2A and (b) 2T3A rasorbers [1].

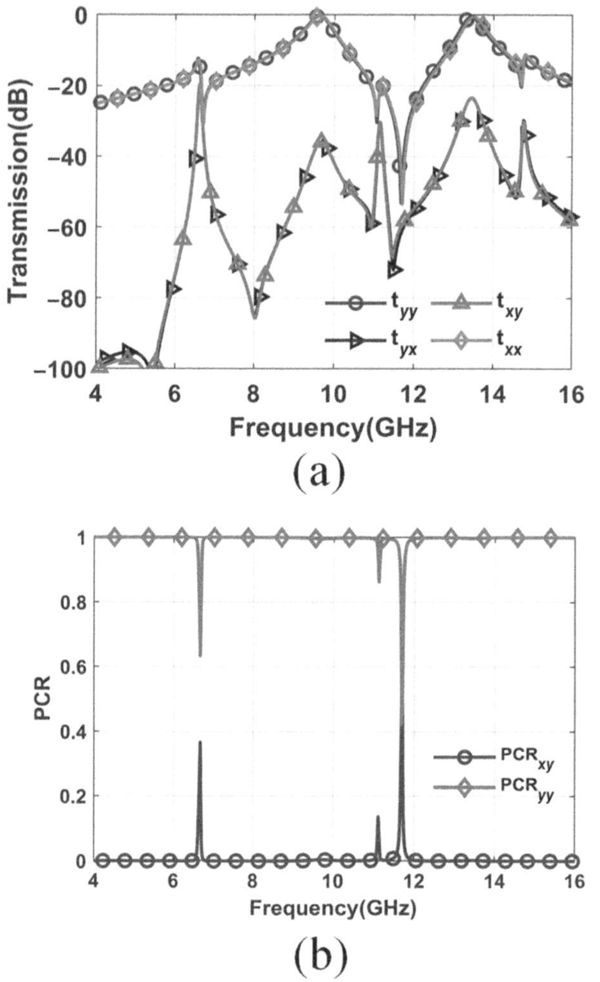

Figure 3.17 (a) Co- and cross-polarization transmission response and (b) PCR of the proposed 2T3A rasorber [1].

3.2 PARAMETRIC STUDIES ON THE PROPOSED RASORBERS

In the proposed 2T2A rasorber, the resonant absorptions flanking the dual transmission bands are attributed to the ring- and cross-shaped metallic resonators positioned on the front side of the substrate. The lower absorption frequency is governed by the ring resonator, whereas the upper absorption frequency is determined by the cross resonator. A detailed parametric study

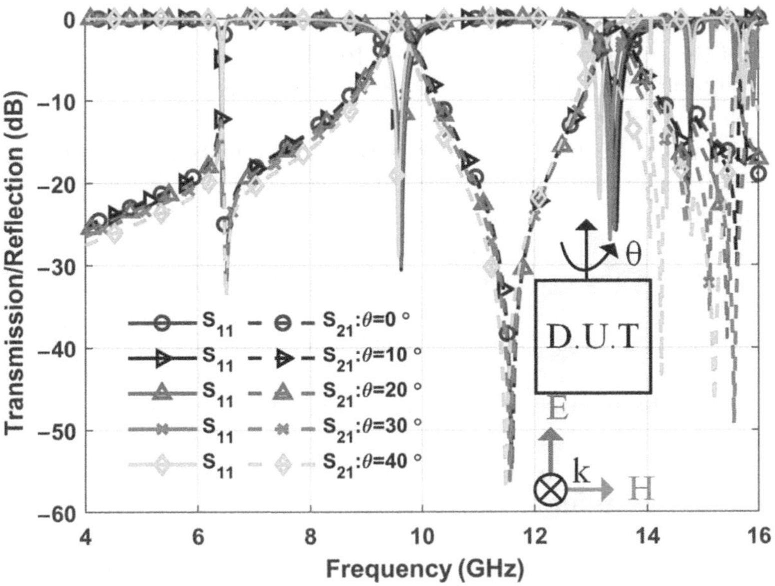

(a)

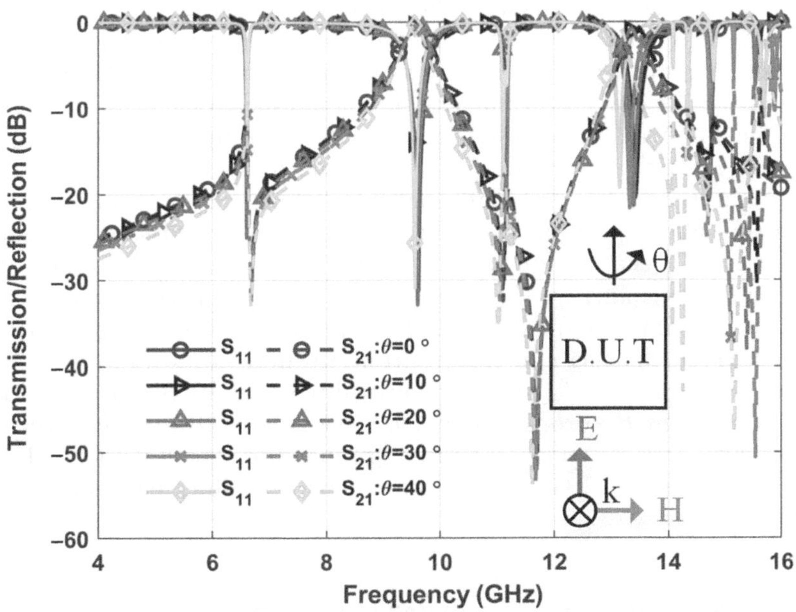

(b)

Figure 3.18 Simulated response under various incident angles for the proposed (a) 2T2A and (b) 2T3A rasorbers [1].

is conducted by varying the geometric dimensions of these resonators while maintaining the fixed bandpass FSS geometry. This analysis enables identification of the tunable frequency ranges for the lower and upper absorption bands within the design framework of the 2T2A rasorber.

The simulated reflection and transmission coefficients for varying values of r_1, while keeping all other parameters constant, are presented in Figure 3.19. The resulting performance metrics of the 2T2A rasorber corresponding to each value of r_1 are summarized in Table 3.1. As evident from the table, reducing r_1 from 6.9 mm to 4.3 mm leads to a significant upward shift in the lower absorption frequency (f_{al}) from 4.61 GHz to 7.51 GHz, whereas the other characteristic frequencies (f_{tl}, f_{tu}, f_{au}) exhibit only minor variations. Notably, when f_{al} exceeds 6.94 GHz, the corresponding absorption drops below 90%. Therefore, to maintain absorption above 90%, the tunable range of f_{al} is limited to 4.61–6.94 GHz, which corresponds to a variation in r_1 from 6.9 mm to 4.7 mm.

In a separate analysis, the reflection and transmission coefficients corresponding to variations in the cross resonator length l_3, while keeping all other parameters fixed, are shown in Figure 3.20. The associated performance metrics at each frequency of interest for different values of l_3 are summarized in Table 3.2. It is observed that reducing l_3 from 7.5 mm to 6.5 mm results in an upward shift of the upper absorption frequency (f_{au}) from 14.02 GHz to 15.86 GHz. The other key frequencies (f_{al}, f_{tl}, f_{tu}) remain largely unaffected. However, to maintain absorption above 90%, the upper absorption frequency f_{au} can be tuned within the range of 14.51–15.07 GHz by adjusting l_3 between 7.1 mm and 6.9 mm, respectively.

In the 2T3A rasorber design, the middle absorption frequency (f_{am}) is determined by the resonance of the split-ring resonator. Therefore, a tunable range for f_{am} can be established by performing a parametric analysis of the split-ring radius r_2. The reflection and transmission coefficients of the 2T3A rasorber for various values of r_2, while keeping all other parameters constant, are shown in Figure 3.21. The performance at each frequency of interest for different r_2 values is summarized in Table 3.3. It is observed that, as r_2 increases from 5.6 mm to 6.6 mm, the middle absorption frequency f_{am} shifts downward from 12.30 GHz to 9.82 GHz. However, when f_{am} drops below 10.37 GHz, the absorption level falls below 90%. Hence, an effective tuning range for f_{am} can be defined from 12.30 GHz to 10.37 GHz, corresponding to r_2 values between 5.6 mm and 6.4 mm.

3.3 EXPERIMENTAL VERIFICATION AND DISCUSSION

Experimental validation of the proposed 2T2A and 2T3A rasorbers was carried out through measurements on their fabricated prototypes. Each prototype comprises an array of 17 × 25 unit cells, fabricated on a 0.254 mm thick

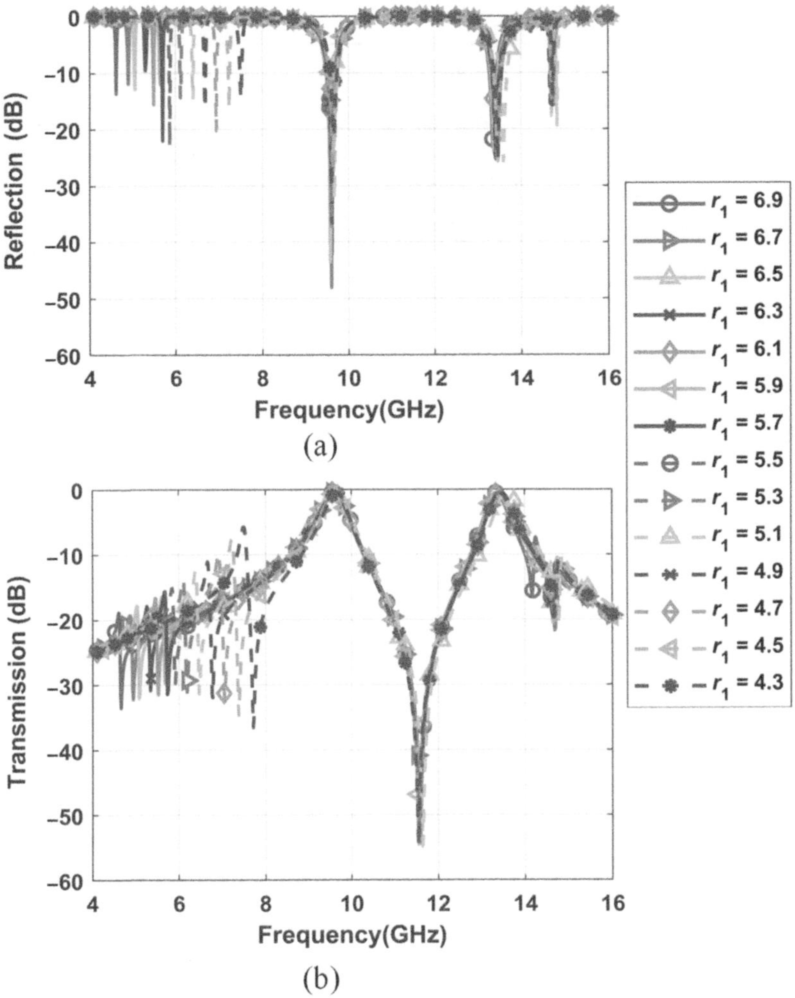

Figure 3.19 Simulated reflection and transmission coefficients of the proposed 2T2A rasorber for varying $\mathbf{r}_1$ (in mm) [1].

Table 3.1 Performance Parameters of 2T2A Rasorber with Varying r_1 [1]

r_l (mm)	f_{al} (GHz)(Ab)	f_{tl} (GHz) (I.L)	f_{tu} (GHz) (I.L)	f_{au} (GHz)(Ab%)
6.9	4.61 (94.64%)	9.61 (0.32 dB)	13.35 (0.53 dB)	14.69 (93.10%)
6.7	4.89 (92.27%)	9.60 (0.28 dB)	13.38 (0.49 dB)	14.74 (92.89%)
6.5	5.05 (94.05%)	9.57 (0.30 dB)	13.42 (0.49 dB)	14.69 (93.33%)
6.3	5.29 (97.70%)	9.60 (0.32 dB)	13.38 (0.50 dB)	14.71 (93.60%)
6.1	5.48 (94.75%)	9.60 (0.30 dB)	13.38 (0.47 dB)	14.76 (92.44%)
5.9	5.64 (92.40%)	9.58 (0.33 dB)	13.36 (0.47 dB)	14.69 (90.84%)
5.7	5.69 (97.32%)	9.61 (0.28 dB)	13.42 (0.49 dB)	14.71 (91.36%)

(Continued)

Table 3.1 (Continued)

r_l (mm)	f_{al} (GHz)(Ab)	f_{tl} (GHz) (I.L)	f_{tu} (GHz) (I.L)	f_{au} (GHz)(Ab%)
5.5	5.86 (97.11%)	9.60 (0.31 dB)	13.41 (0.47 dB)	14.69 (91.00%)
5.3	6.12 (92.84%)	9.61 (0.27 dB)	13.42 (0.47 dB)	14.71 (92.71%)
5.1	6.41 (92.57%)	9.67 (0.28 dB)	13.59 (0.45 dB)	14.82 (92.36%)
4.9	6.69 (92.13%)	9.61 (0.28 dB)	13.42 (0.49 dB)	14.67 (92.70%)
4.7	6.94 (91.61%)	9.63 (0.30 dB)	13.46 (0.47 dB)	14.69 (92.74%)
4.5	7.23 (81.45%)	9.65 (0.30 dB)	13.45 (0.46 dB)	14.71 (90.56%)
4.3	7.51 (71.67%)	9.65 (0.36 dB)	13.46 (0.50 dB)	14.72 (91.92%)

Note: Ab: Absorption; I.L.: insertion loss; f_{al} : lower absorption frequency; f_{tl}: lower transmission frequency; f_{tu} : upper transmission frequency; f_{au} : upper absorption frequency.

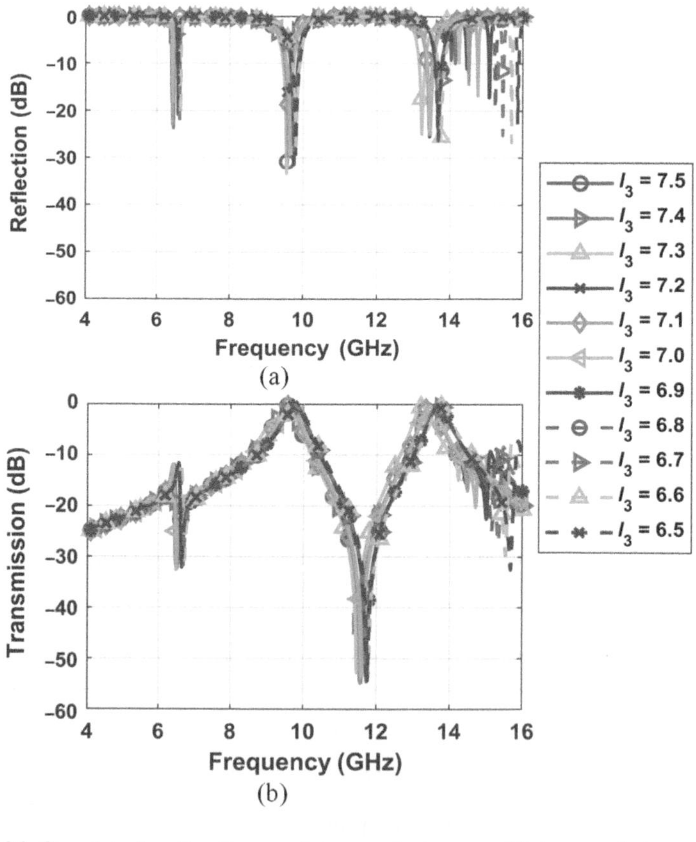

Figure 3.20 Simulated reflection and transmission coefficients of the proposed 2T2A rasorber for varying l_3 (in mm) [1].

Table 3.2 Performance Parameters of 2T2A Rasorber with Varying l_3 [1]

$l_3\,(mm)$	$f_{al}\,(GHz)(Ab)$	f_{tl} (GHz) (I.L)	f_{tu} (GHz) (I.L)	$f_{au}\,(GHz)(Ab\%)$
7.5	6.45 (91.25%)	9.56 (0.29 dB)	13.43 (0.51 dB)	14.02 (78.79%)
7.4	6.45 (94.48%)	9.56 (0.28 dB)	13.45 (0.47 dB)	14.13 (82.98%)
7.3	6.42 (92.57%)	9.54 (0.28 dB)	13.22 (0.51 dB)	14.25 (85.90%)
7.2	6.41 (92.97%)	9.60 (0.38 dB)	13.43 (0.47 dB)	14.46 (88.74%)
7.1	6.42 (94.52%)	9.58 (0.30 dB)	13.43 (0.48 dB)	14.51 (92.29%)
7.0	6.43 (93.62%)	9.63 (0.29 dB)	13.46 (0.48 dB)	14.76 (92.29%)
6.9	6.57 (93.29%)	9.75 (0.27 dB)	13.69 (0.43 dB)	15.07 (90.61%)
6.8	6.57 (92.35%)	9.77 (0.27 dB)	13.67 (0.43 dB)	15.24 (89.62%)
6.7	6.62 (94.17%)	9.75 (0.30 dB)	13.70 (0.42 dB)	15.45 (88.02%)
6.6	6.64 (93.00%)	9.79 (0.28 dB)	13.74 (0.41 dB)	15.69 (83.80%)
6.5	6.57 (90.80%)	9.79 (0.27 dB)	13.67 (0.42 dB)	15.86 (82.55%)

Note: Ab: absorption; I.L.: insertion loss; f_{al}: lower absorption frequency; f_{tl}: lower transmission frequency; f_{tu}: upper transmission frequency; f_{au}: upper absorption frequency.

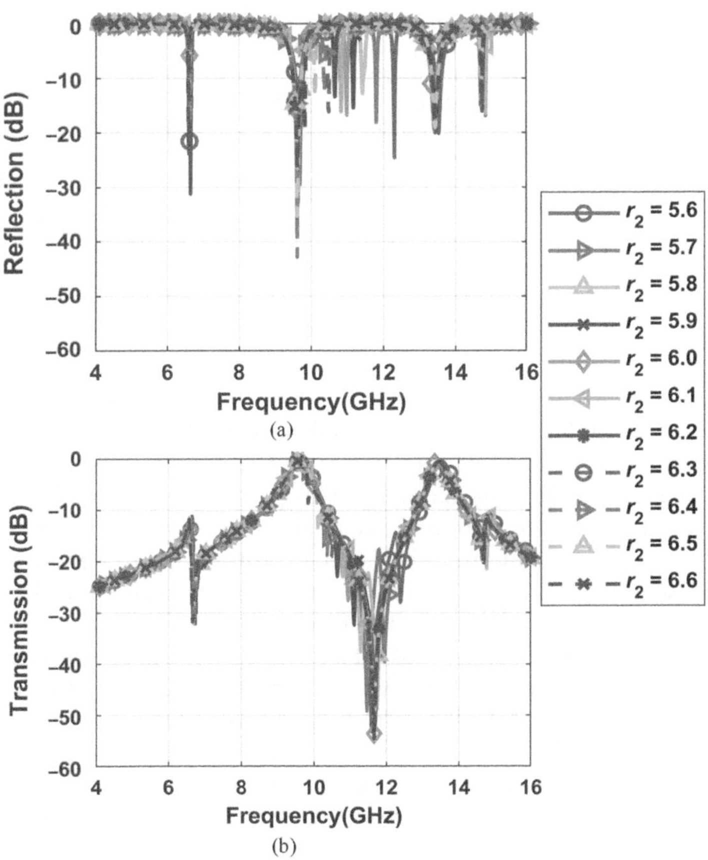

Figure 3.21 Simulated reflection and transmission coefficients of the proposed 2T3A rasorber for varying r_2 (in mm) [1].

Table 3.3 Performance Parameters of 2T3A Rasorber with Varying r_2 [1]

r_2 (mm)	f_{al} (GHz) (Ab%)	f_{tl} (GHz) (I.L)	f_{am} (GHz) (Ab%)	f_{tu} (GHz) (I.L)	f_{au} (GHz) (Ab%)
5.6	6.62 (95.08%)	9.68 (0.29 dB)	12.30 (97.06%)	13.53 (0.49 dB)	14.84 (91.69%)
5.7	6.66 (93.01%)	9.61 (0.29 dB)	11.80 (96.55%)	13.44 (0.55 dB)	14.70 (91.89%)
5.8	6.67 (91.06%)	9.63 (0.29 dB)	11.42 (93.87%)	13.41 (0.63 dB)	14.74 (92.67%)
5.9	6.60 (94.00%)	9.63 (0.31 dB)	11.17 (96.81%)	13.43 (0.52 dB)	14.75 (91.91%)
6.0	6.64 (92.74%)	9.61 (0.28 dB)	10.98 (97.25%)	13.41 (0.58 dB)	14.74 (91.83%)
6.1	6.66 (93.58%)	9.64 (0.32 dB)	10.83 (96.94%)	13.42 (0.52 dB)	14.83 (92.83%)
6.2	6.64 (94.71%)	9.61 (0.28 dB)	10.66 (94.07%)	13.45 (0.50 dB)	14.74 (92.73%)
6.3	6.67 (91.26%)	9.61 (0.29 dB)	10.50 (95.53%)	13.42 (0.52 dB)	14.74 (91.02%)
6.4	6.63 (93.18%)	9.61 (0.30 dB)	10.37 (92.74%)	13.45 (0.52 dB)	14.69 (92.69%)
6.5	6.60 (92.45%)	9.61 (0.28 dB)	10.10 (87.54%)	13.44 (0.51 dB)	14.71 (92.22%)
6.6	6.59 (93.53%)	9.61 (0.34 dB)	9.82 (54.94%)	13.45 (0.56 dB)	14.72 (92.73%)

Note: Ab: absorption; I.L.: insertion loss; f_{al} : lower absorption frequency; f_{tl}: lower transmission frequency; f_{tu} : upper transmission frequency; f_{am} : middle absorption frequency; f_{au} : upper absorption frequency.

Taconic substrate with a dielectric constant (ε_r) of 2.2 and a loss tangent (tanδ) of 0.0009. Photographs of the fabricated 2T2A and 2T3A rasorber prototypes, each with overall dimensions of 265 mm × 385 mm, are presented in Figure 3.22. The bandpass FSS pattern on the back side of the substrate is identical for both designs, as shown in Figure 3.22(c).

The measurements of the fabricated prototypes were conducted in an anechoic chamber using a free-space measurement technique. Three pairs of standard gain horn antennas, covering the J (5.85–8.20 GHz), X (8.20–12.40 GHz), and Ku (12.40–18.0 GHz) bands, were used sequentially and connected to a Keysight PNA Network Analyzer (N5224B). The experimental setups for reflection and transmission measurements are illustrated in Figures 3.23(a) and 3.23(b), respectively. Initially, reference measurements were performed in both setups to enable normalization of the measured data and obtain effective reflection and transmission coefficients. For reflection, the reference was a metallic sheet, while for transmission, the reference was the direct free-space path loss between the two antennas. Figure 3.24

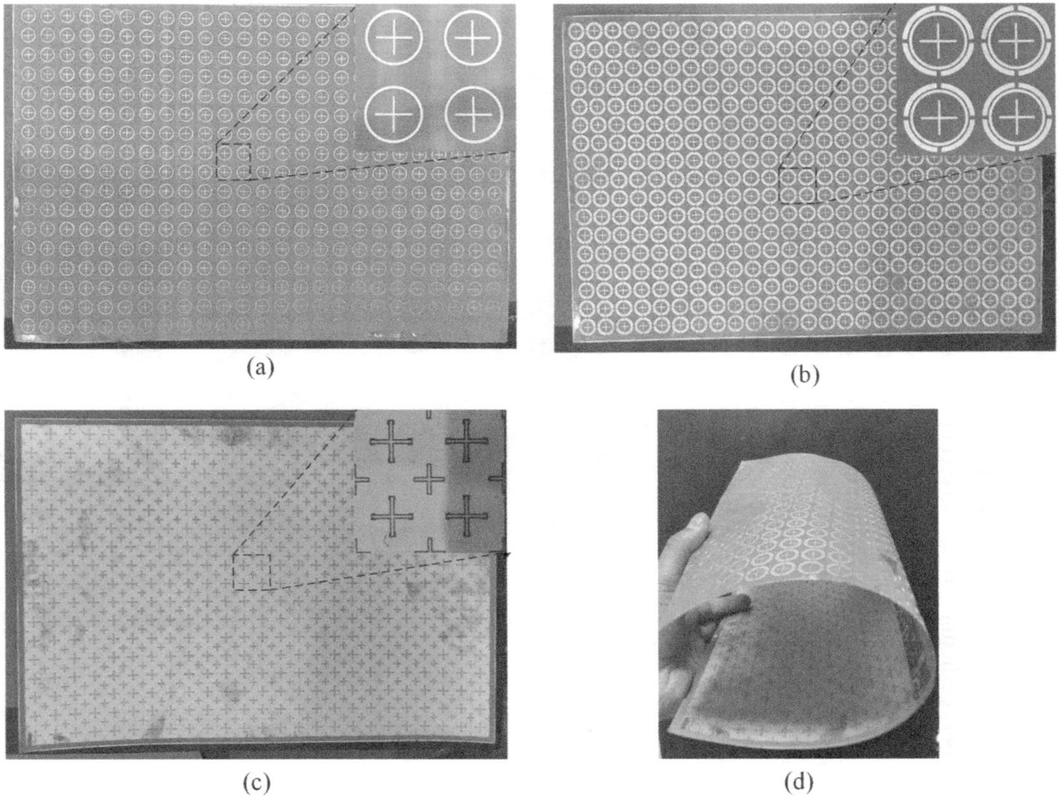

Figure 3.22 Photograph of the fabricated prototypes of the proposed rasorbers: (a) front side of 2T2A rasorber, (b) front side of 2T3A rasorber, (c) back side of both 2T2A and 2T3A rasorbers, and (d) flexible view of 2T3A rasorber [1].

presents a comparison between the simulated and measured reflection and transmission coefficients for both the 2T2A and 2T3A rasorbers. The close match between simulated and experimental results confirms the validity of the proposed rasorber designs.

The performance of the proposed rasorbers under oblique incidence was experimentally evaluated. The schematic diagrams of the measurement setups for capturing reflection and transmission coefficients at oblique angles are shown in Figure 3.25. During the measurement, the rasorber structure (Device Under Test–D.U.T.) is positioned at the center of a defined circular arc. The direction perpendicular to the rasorber surface represents the normal incidence (0° line). To simulate oblique incidence, the transmitting and receiving antennas are rotated along the circular arc, either clockwise or counterclockwise, so as to form the desired angles with respect to the normal direction.

For reflection measurements, as illustrated in Figure 3.25(a), both the transmitting and receiving antennas are positioned on the front side of the structure. Under normal incidence, the antennas are aligned along the 0° line. For oblique incidence, the transmitting and receiving antennas are rotated in opposite directions, one clockwise and the other counterclockwise, along the

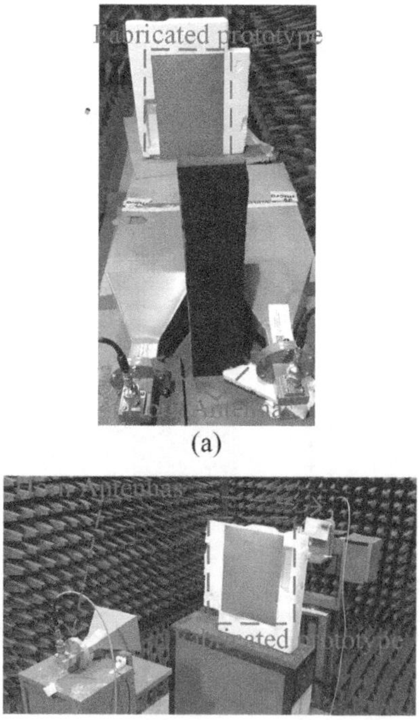

Figure 3.23 Photograph of the experimental setup for (a) reflection and (b) transmission measurements [1].

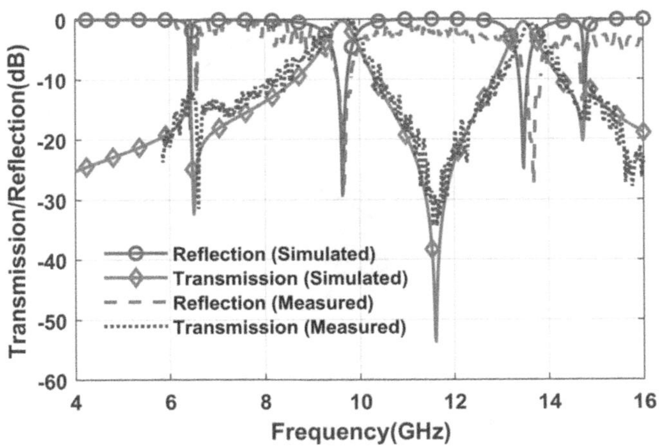

(a)

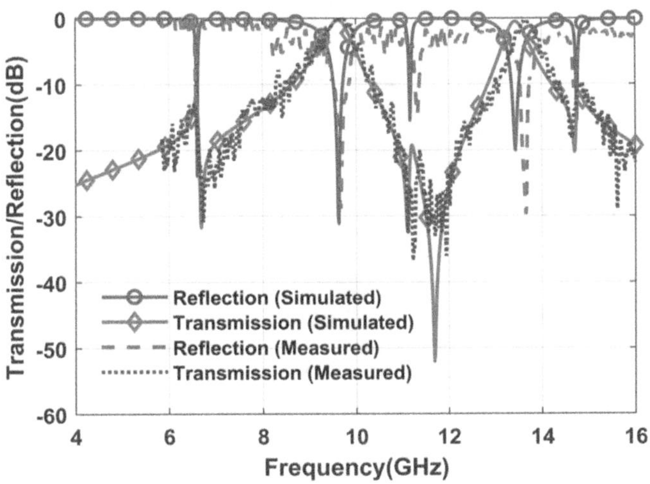

(b)

Figure 3.24 Simulated and measured reflection and transmission coefficients of the proposed (a) 2T2A and (b) 2T3A rasorbers [1].

circular arc, forming equal angles with respect to the normal. This measurement approach adheres to the fundamental principle of electromagnetic wave propagation, where the angle of incidence equals the angle of reflection.

For measuring the transmission coefficient, the transmitting antenna is positioned in front of the structure, while the receiving antenna is placed behind it, as shown in Figure 3.25(b). Under normal incidence, both antennas are aligned along the 0° axis. For oblique incidence, the antennas are moved along

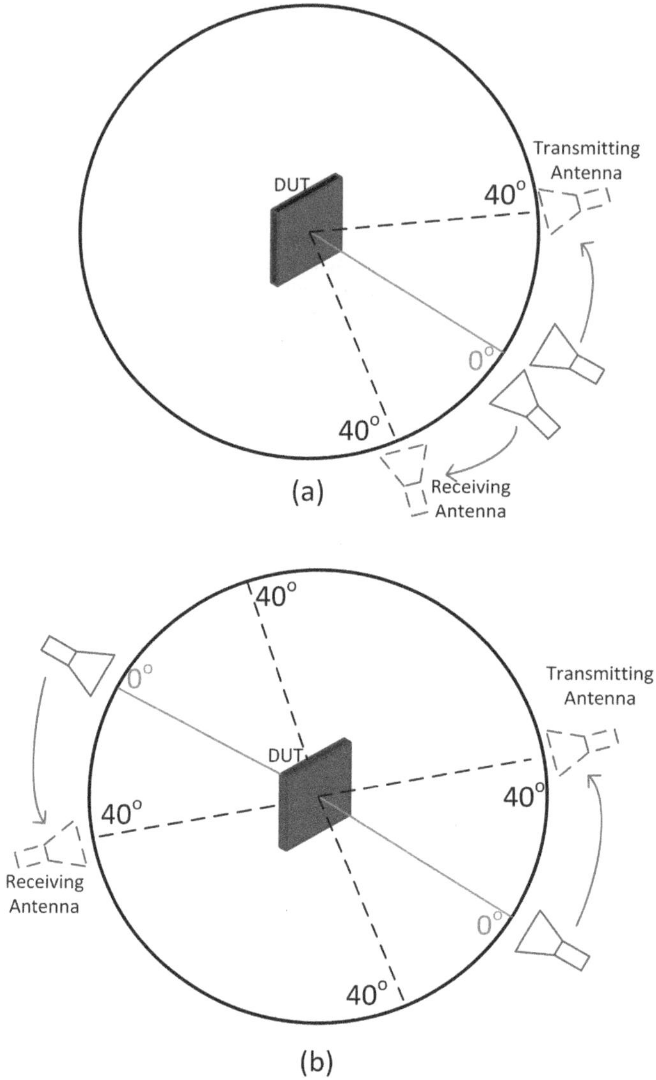

Figure 3.25 Schematic of the setups for the (a) reflection and (b) transmission measurement of the proposed rasorber (D.U.T.) under oblique incidence [1].

the circular path in the same direction (either clockwise or counterclockwise), forming the desired oblique angle relative to the normal incidence path.

The measured reflection and transmission coefficients of the rasorber prototypes under oblique incidence for TE polarization are presented in Figure 3.26. The results confirm that the proposed rasorbers maintain

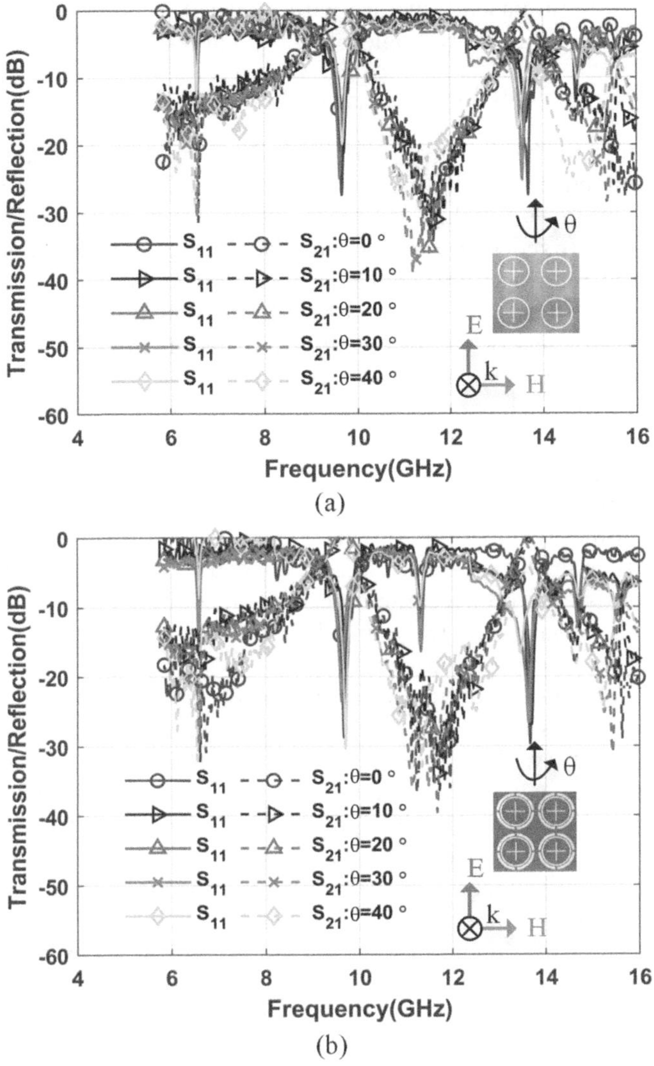

Figure 3.26 Measured reflection and transmission coefficients under oblique incident angles (TE polarized incident wave) for the proposed (a) 2T2A and (b) 2T3A rasorbers [1].

stable performance up to an incident angle of 40°, thereby experimentally validating their angular stability.

To validate the conformal behavior of the proposed rasorber, reflection and transmission measurements for a TE-polarized incident wave were also conducted on a curved surface. The fabricated prototype was mounted on a cylindrical, foam-based structure with a curvature applied along its narrower side and a curvature angle of 120°. The radius r of the cylindrical structure was calculated as $r = \frac{180A}{\pi D_C}$, where A is the arc length (265 mm, corresponding to the narrower side) and D_C is the curvature angle. Photographs of the curved configurations for both the 2T2A and 2T3A rasorbers are shown in Figure 3.27. The reflection reference was obtained using a metal sheet affixed to the same cylindrical foam. Figure 3.28 compares the measured reflection and transmission coefficients of the curved structure with those of the flat version for both rasorber types. The close match between the two sets of measurements confirms the flexible and conformal performance of the proposed rasorbers.

The conformal behavior of the proposed rasorber structures is further evaluated by increasing the degree of curvature D_C to 150° and 180°. To achieve these curvatures, the prototypes were mounted on cylindrical foam structures with radii of 101.2 mm and 84.3 mm, respectively. Figure 3.29 presents a comparison of the measured responses of the curved 2T2A and 2T3A rasorber prototypes for different curvature angles. The results demonstrate consistent and acceptable performance at all frequencies of interest, even at a curvature of 180°, thereby confirming the superior conformal capability of the proposed designs.

Table 3.4 presents a performance comparison between the proposed ultra-thin, conformal rasorbers and previously reported multiband absorbers and wideband rasorbers. Unlike the multiband absorbers in [3–5] the proposed designs offer enhanced functionality by achieving dual transmission bands in addition to dual or triple absorption bands. Compared to the wideband absorptive rasorbers reported in [6–12], the proposed structures are ultra-thin, conformal, and implemented on a single-layer substrate without the need for any external lumped components. Moreover, their flexibility is experimentally validated, demonstrating conformal operation even at a curvature of up to 180°, confirming their suitability for non-planar applications.

3.4 CONCLUSION

In this chapter, a single-layer flexible rasorber structure (A-T-T-A/A-T-A-T-A) is proposed, featuring dual transmission bands along with dual or triple absorption bands. The design integrates a dual-band bandpass FSS with a dual-band resonant absorber such that, at the absorption frequencies, the bandpass FSS exhibits maximum reflection. This configuration results in the realization of a 2T2A rasorber, where two absorption bands (f_{al} = 6.4 GHz and f_{au} = 14.7 GHz) are positioned on either side of the transmission bands (f_{tl} = 9.6 GHz and f_{tu} = 13.4 GHz). The structure is further modified to form a 2T3A rasorber by introducing an additional absorption band (f_{am} = 11.1 GHz) between the two transmission bands. Prototypes comprising 17 × 25 unit cells of each rasorber design are fabricated and experimentally validated. Additionally, the conformal nature of the proposed rasorbers is confirmed through measurements on curved prototypes with varying curvature angles.

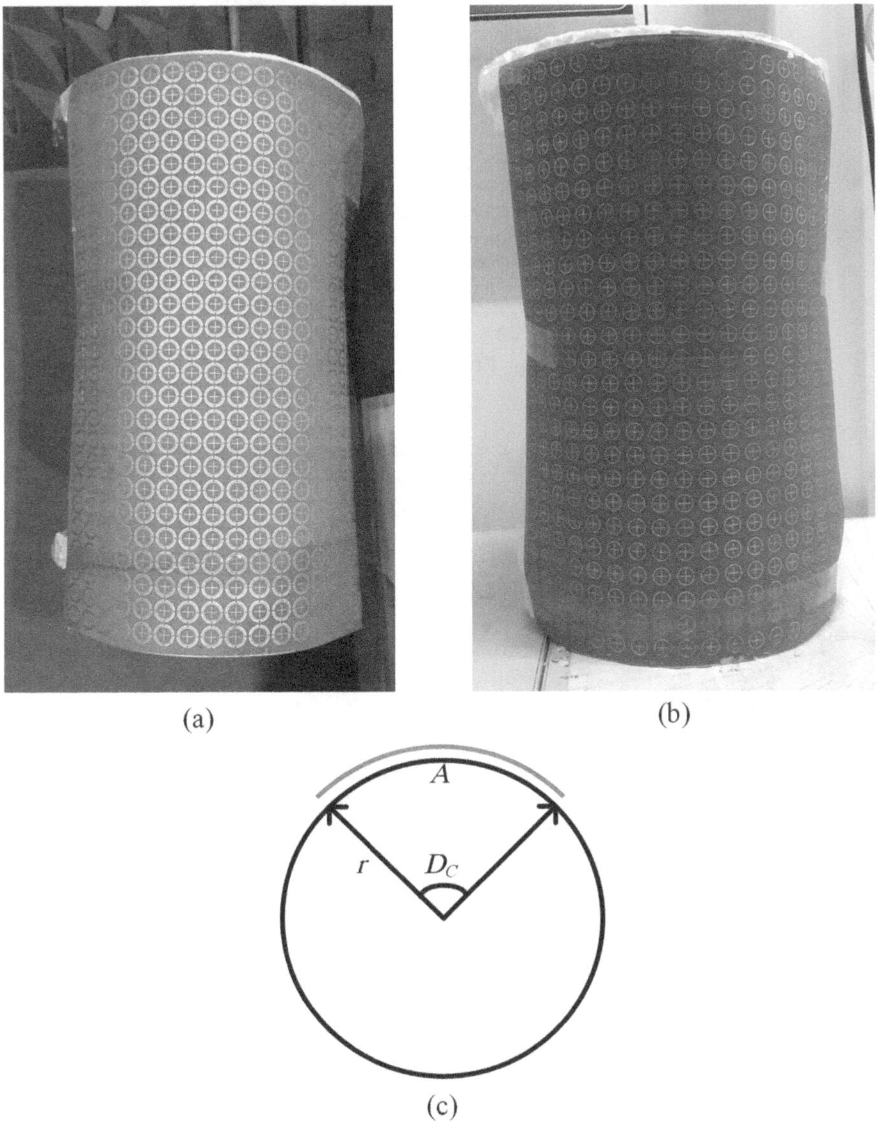

Figure 3.27 Photograph of the curved prototype of the proposed (a) 2T2A and (b) 2T3A rasorbers. (c) Lateral schematic of the curved prototype with radius *r*, degree of curvature D_C, and arc length *A* [1].

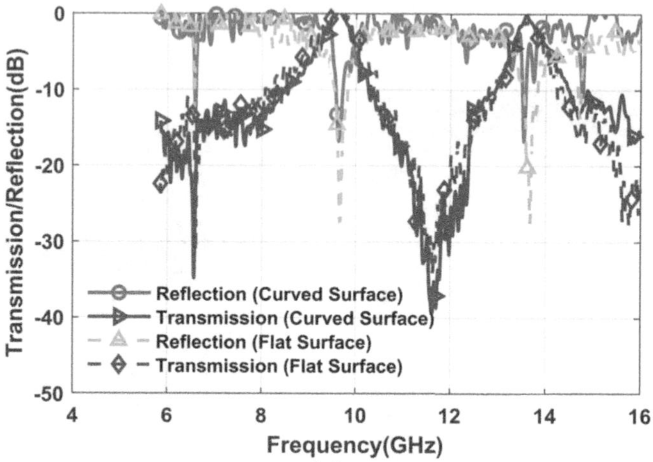

(a)

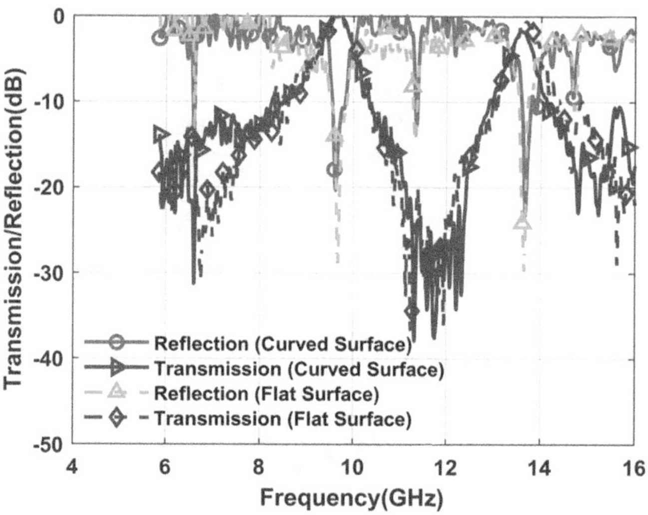

(b)

Figure 3.28 Comparison of the measured reflection and transmission coefficients for flat and curved surfaces with degree of curvature D_C = 120° for the proposed (a) 2T2A and (b) 2T3A rasorbers under the normal TE incidence [1].

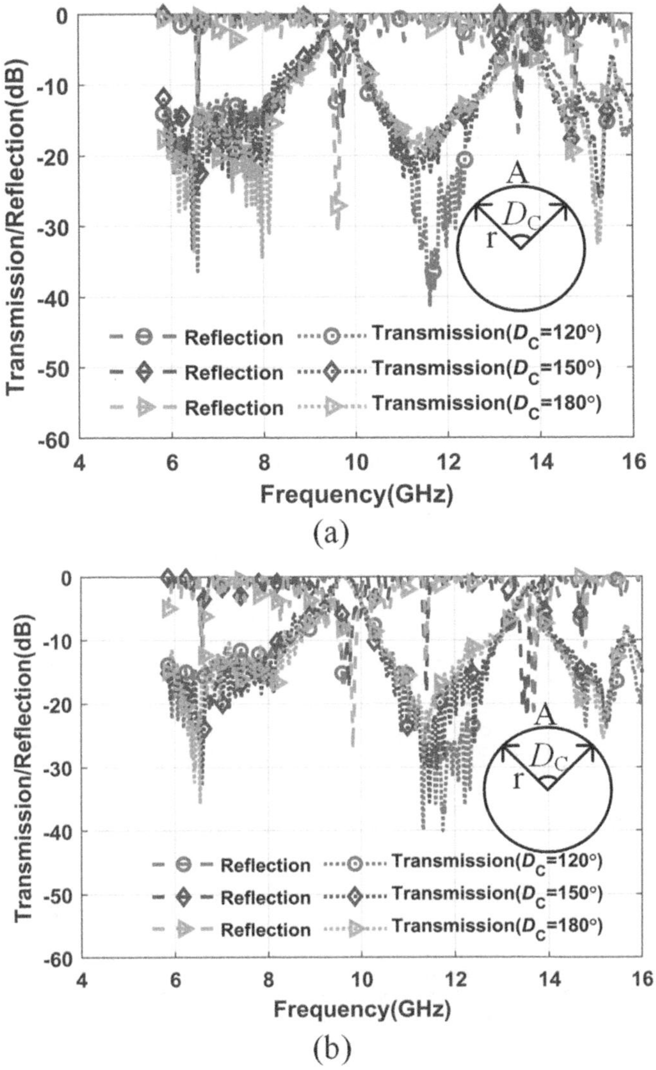

Figure 3.29 Comparison of the measured reflection and transmission coefficients for curved surface with various degree of curvature (D_C) for the proposed (a) 2T2A and (b) 2T3A rasorbers under the normal TE incidence [1].

Table 3.4 Performance Comparison of Proposed FSS-Based Rasorber with Earlier Reported Multi/Wideband Absorbers/Rasorbers [1]

Ref.	*Absorption*	*Transmission/Min. Insertion Loss*	*Thick*	*Size*	*N.O.L*	*L.E*	*Polarization*	*Con.*
[3]	Dual-band (16.6 and 24.4 GHz)	Nil	$0.022\lambda_0$	$0.076\lambda_0^2$	1	No	Insensitive	No
[4]	Penta-band (9.9, 10.4, 10.9, 11.7, and 13.2 GHz)	Nil	$0.026\lambda_0$	$0.146\lambda_0^2$	1	No	Sensitive	No
[5]	Triple-band (4.1, 6.6, 9.9 GHz)	Nil	$0.007\lambda_0$	$0.038\lambda_0$	1	No	Insensitive	Yes
[6]	Wideband (4.0 - 6.2 GHz, 8.2–12.0) GHz	Dual-band (7.2 GHz / 2.3 dB and 13.0 GHz / 1.69 dB)	$0.107\lambda_0$	$0.029\lambda_0^2$	2	Yes	Sensitive	No
[7]	Wideband (1.6 - 2.9 GHz, 3.8 - 4.7GHz, 5.3 - 6.9GHz)	Dual-band (3.5 GHz/0.25 dB and 4.9 GHz / 0.1 dB)	$0.076\lambda_0$	$0.021\lambda_0^2$	2	Yes	Dual	No
[8]	Wideband (3.8 - 10.8 GHz)	Low-pass (< 860 MHz)/ (<1 dB)	$0.140\lambda_0$	$0.048\lambda_0^2$	2	Yes	Dual	No
[9]	Wide band (1.5 - 4.6 GHz, 9.2 - 13.7 GHz)	6.5 GHz / 0.67 dB	$0.171\lambda_0$	$0.010\lambda_0^2$	4	Yes	Dual	No
[10]	Wide band (2.3 - 5.3 GHz, 7.8 - 14.6 GHz)	6.3 GHz / 0.78 dB	$0.111\lambda_0$	$0.023\lambda_0^2$	3	Yes	Dual	No
[11]	Wide band (2.5 - 4.6 GHz, 7.7 - 12.0 GHz)	6.0 GHz/0.27 dB	$0.075\lambda_0$	$0.010\lambda_0^2$	2	Yes	Dual	No
[12]	Wideband (2.9–5.6 GHz, 7.4–9.2 GHz)	Single-band (6.0 GHz/0.38 dB)	$0.0639\lambda_0$	$0.038\lambda_0^2$	1	Yes	Dual	No
	Wideband (3.1 - 5.0GHz, 6.4 - 8.3GHz)	Single-band (5.8 GHz/0.45 dB)	$0.0878\lambda_0$	$0.042\lambda_0^2$	1	Yes	Dual	No
	Wideband (3.2–6.4 GHz, 8.3–10.5 GHz)	Single-band (7.7GHz / 0.25 dB)	$0.0987\lambda_0$	$0.045\lambda_0^2$	2	Yes	Dual	Yes
This work	Dual-band (6.4 and 14.7 GHz)	Dual-band (9.6 GHz/0.29 dB and 13.4 GHz/0.48 dB)	$0.0054\lambda_0$	$0.101\lambda_0^2$	1	No	Insensitive	Yes
	Triple-band (6.6, 11.1 and 14.7 GHz)	Dual-band (9.6 GHz/0.31 dB and 13.4 GHz/0.52 dB)	$0.0055\lambda_0$	$0.108\lambda_0^2$	1	No	Insensitive	Yes

Note: λ_0 = free space wavelength at the lowest frequency of interest; Con. = conformal; L.E. = lumped elements; N.O.L. = no. of layers.

REFERENCES

1. M. M. Zargar et al., "Single-Layered Flexible Dual Transmissive Rasorbers with Dual/Triple Absorption Bands for Conformal Applications," IEEE Access, vol. 9, pp. 150426–150442, 2021.
2. S. Ghosh and K. V. Srivastava, "An Equivalent Circuit Model of FSS-Based Metamaterial Absorber Using Coupled Line Theory," IEEE Antenn. Wirel. Propag. Lett., vol. 14, pp. 511–514, 2015.
3. J. Wang, R. Yang, J. Tian, X. Chen and W. Zhang, "A Dual-Band Absorber with Wide-Angle and Polarization Insensitivity," IEEE Antenn. Wirel. Propag. Lett., vol. 17, no. 7, pp. 1242–1246, Jul. 2018.
4. P. Jain, A. K. Singh, J. K. Pandey, S. Garg, S. Bansal, M. Agarwal, S. Kumar, N. Sardana, N. Gupta and A. K. Singh, "Ultra-Thin Metamaterial Perfect Absorbers for Single-/Dual-/Multi-Band Microwave Applications," IET Microw. Antenn. Propag., vol. 14, no. 5, pp. 390–396, Apr. 2020.
5. A. K. Singh, M. P. Abegaonkar and S. K. Koul, "Dual- and Triple-Band Polarization Insensitive Ultrathin Conformal Metamaterial Absorbers With Wide Angular Stability," IEEE Trans. Electromagn. Compat., vol. 61, no. 3, pp. 878–886, Jun. 2019.
6. M. Guo, Q. Chen, Z. Sun, D. Sang and Y. Fu, "Design of Dual-Band Frequency-Selective Rasorber," IEEE Antenn. Wirel. Propag. Lett., vol. 18, no. 5, pp. 841–845, May 2019.
7. X. Xiu, W. Che, W. Yang, Y. Han and Q. Xue, "Double-Polarized Dual Passband Absorptive Frequency-Selective Transmission Structure," IEEE Trans. Electromagn. Compat., vol. 62, no. 5, pp. 1951–1960, Oct. 2020.
8. G. Q. Luo, W. Yu, Y. Yu, H. Jin, K. Fan and F. Zhu, "Broadband Dual Polarized Band-Absorptive Frequency-Selective Rasorber Using Absorptive Transmission/Reflection Surface," IEEE Trans. Antenn. Propag., vol. 68, no. 12, pp. 7969–7977, Dec. 2020.
9. H. Ye, J. Wei, L. Lin, F. Liu, L. Miao, S. Bie and J. Jiang, "A Frequency Selective Surface Rasorber Based on Four Functional Layers," IEEE Trans. Antenn. Propag., vol. 69, no. 5, pp. 2768–2778, May 2021.
10. J. Xia et al., "Design of a Wideband Absorption Frequency Selective Rasorber Based on Double Lossy Layers," IEEE Trans. Antenn. Propag., vol. 68, no. 7, pp. 5718–5723, Jul. 2020.
11. Q. Guo, J. Su, Z. Li, J. Song and Y. Guan, "Miniaturized-Element Frequency selective Rasorber Design Using Characteristic Modes Analysis," IEEE Trans. Antenn. Propag., vol. 68, no. 9, pp. 6683–6694, Sept. 2020.
12. S. C. Bakshi, D. Mitra and F. L. Teixeira, "Wide-Angle Broadband Rasorber for Switchable and Conformal Application," IEEE Trans. Microw. Theory Techn., vol. 69, no. 2, pp. 1205–1216, Feb. 2021.

Chapter 4

A-T FSR Using Modified Double Cross-Slot Bandpass Layer

In the earlier chapters, thin, single-layer rasorber structures demonstrating narrowband absorption characteristics were discussed. Starting with this chapter, the focus shifts toward the development of rasorbers that support broadband absorption. Unlike the narrowband configurations, the broadband rasorber designs utilize multilayered architectures integrated with lossy lumped elements. This chapter introduces a polarization-insensitive A-T frequency selective surface (FSR), where the concepts of wideband absorption and passband transmission are first explored independently and later integrated to achieve the desired absorptive-transmissive behavior. The designed FSR delivers a broad absorption range coupled with a transmission band located at a higher frequency. It maintains stable performance for incident angles up to 30°. The operation of the FSR is examined through an equivalent circuit model (ECM), and its effectiveness is demonstrated through the fabrication and experimental validation of a prototype.

4.1 DESIGN AND ANALYSIS OF THE FSR

A frequency selective rasorber (FSR) is generally composed of multiple layers, typically including both lossy and lossless components. Its design methodology involves integrating two distinct functionalities of frequency selective surfaces (FSS): broadband electromagnetic absorption and selective bandpass transmission. To realize the FSR, the conventional metallic backing used in absorbers is substituted with a bandpass FSS. This FSS must behave like a perfect reflector (i.e., a metallic ground) within the absorption frequency range. The development of the FSR is a three-stage process. Initially, a wideband absorber is designed to ensure strong absorption over the target frequency range. The second stage focuses on the design and refinement of the bandpass FSS by tuning its various parameters. In the final stage, the broadband absorber and the optimized bandpass FSS are combined, and the overall structure is fine-tuned to achieve the desired FSR performance.

DOI: 10.1201/9781003644323-4

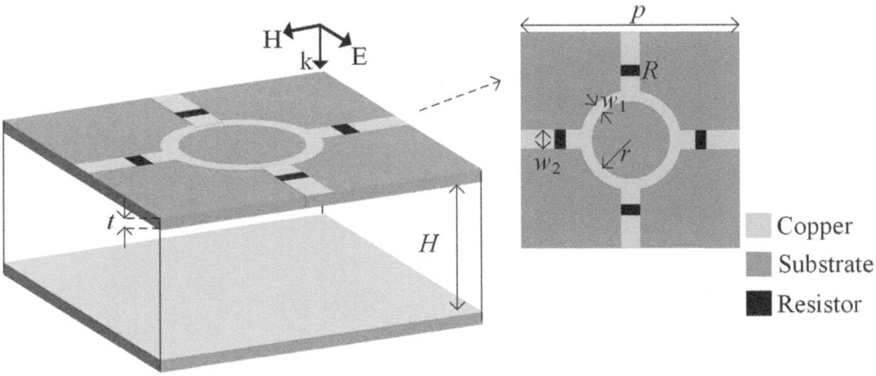

Figure 4.1 Unit cell schematic of the broadband absorber [1].

4.1.1 Analysis of the Broadband Absorber

The broadband absorber's unit cell is built using a two-layer configuration on an FR-4 substrate with a thickness of 0.8 mm, a relative permittivity of $\epsilon_r = 4.4$, and a loss tangent of tanδ = 0.02. A schematic of the unit cell is illustrated in Figure 4.1. On the front face of the upper layer, a circular cross-shaped resonator is printed, while the back side of this layer is fully etched. Chip resistors (R) are inserted in the gaps of the cross resonator structure. The bottom layer comprises a copper-clad substrate that serves as a ground plane. An air gap of height *H* is introduced between the two layers to form the desired absorber configuration. The absorber's unit cell is modeled and analyzed using the CST electromagnetic simulator, and the reflection response for various resistor values and air gap heights is plotted in Figure 4.2. The optimal resistor value that provides enhanced reflection bandwidth is found to be 200 ohms. The best performance is achieved when *H* is set to 12 mm, which corresponds to a quarter-wavelength at the center of the absorption band. The final optimized design parameters are $p = 20$, $r = 5$, $w_1 = 0.5$, $w_2 = 1$, and $t = 0.8$ mm. The resulting structure achieves effective broadband absorption from approximately 3.2 GHz to 8.9 GHz, where the reflection coefficient remains below –10 dB across the entire range.

4.1.2 Analysis of the Bandpass FSS

To introduce a transmission band located at a higher frequency than the broadband absorption range discussed earlier, a bandpass frequency selective surface (FSS) is developed. The unit cell of this FSS is designed to align dimensionally with that of the circular cross-shaped broadband absorber. As illustrated in Figure 4.3(a), the unit cell incorporates a double cross-slot (DCS) etched into a metallic substrate. These etched slots allow specific frequency components of the incident electromagnetic wave to pass through,

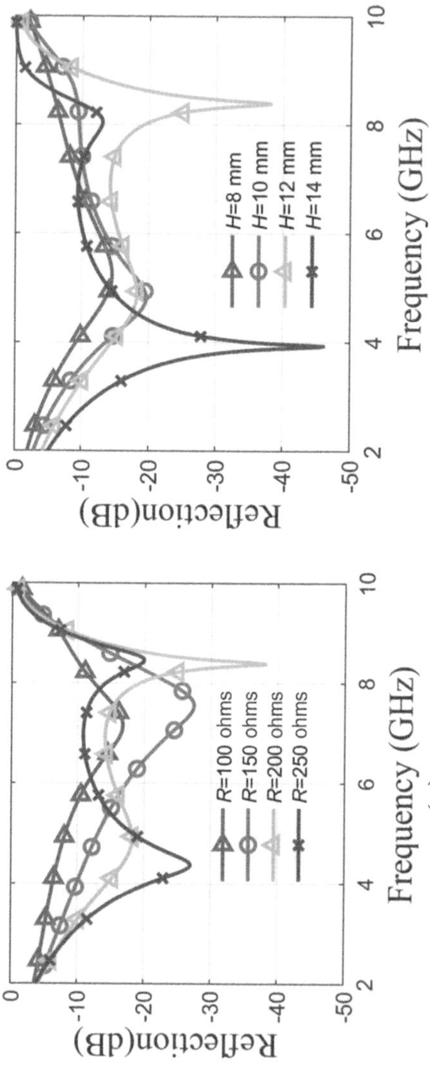

Figure 4.2 Simulated reflection coefficient of broadband absorber with varying (a) **R** and (b) **H** [1].

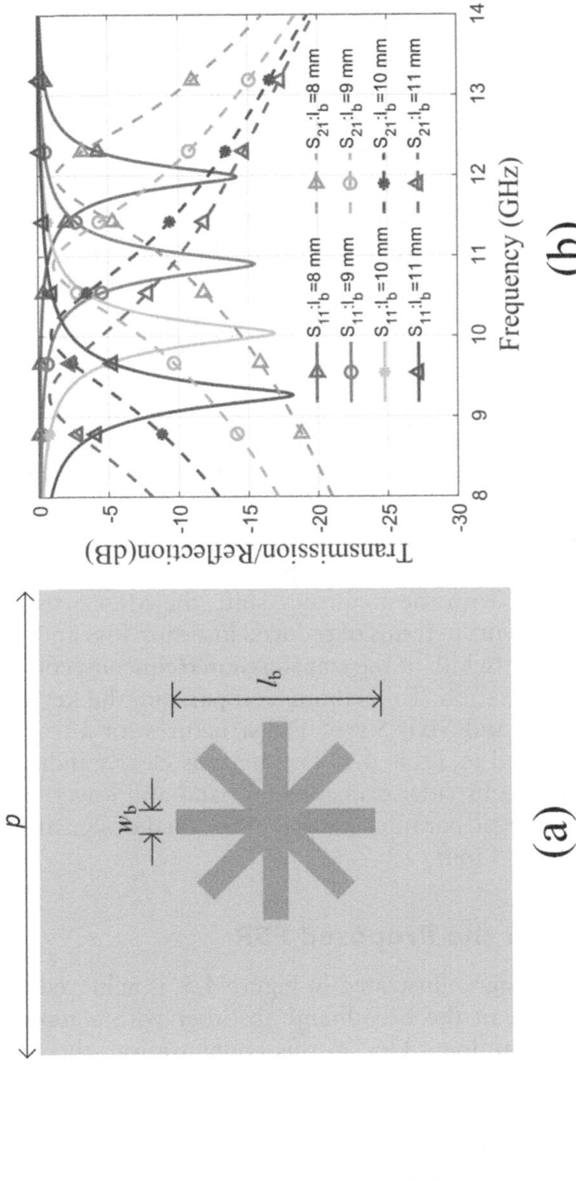

Figure 4.3 DCS-shaped bandpass frequency selective surface: (a) unit cell schematic and (b) simulated response [1].

functioning as a bandpass filter centered at a particular resonance. The transmission frequency is inversely related to the length of the slots – shorter slots correspond to higher resonance frequencies. Figure 4.3(b) presents the simulated transmission characteristics for varying slot lengths l_b. The goal is to ensure that the FSS maintains a passband at a higher frequency, while behaving like a solid metallic reflector across the lower absorption band from 3.2 GHz to 8.9 GHz. From the simulation results, it is evident that, as the slot length decreases, the center frequency of the passband shifts upward. However, a trade-off is observed: reducing the slot length also leads to higher insertion loss at the corresponding resonance. This degradation occurs because shorter slots result in a lower effective aperture within the unit cell, thereby weakening transmission efficiency. Therefore, for optimal filtering performance and reduced insertion loss, the slot dimensions must be appropriately scaled relative to the size of the unit cell.

To enhance the transmission characteristics of the bandpass FSS at elevated frequencies, the original FSS design is refined by increasing the number of resonant slots within each unit cell. This higher slot density is achieved by incorporating a replicated quarter-section of the double cross-slot (DCS) pattern at each corner of the unit cell, as illustrated in Figure 4.4(a). The modified configuration, referred to as the modified double cross-slot (MDCS) structure, is simulated for different slot lengths l_b, with the results presented in Figure 4.4(b). In the MDCS design, the introduction of additional slots alters the effective inductance and capacitance of the structure, resulting in a shift of the resonance frequency toward higher values for the same slot size. Along with the frequency shift, the MDCS structure offers significant improvements in terms of reduced insertion loss and an expanded −10 dB reflection bandwidth at the transmission frequency, compared to the original DCS configuration. To facilitate comparison, the key performance metrics of both DCS and MDCS-type FSS structures for a fixed slot length of l_b = 9 mm are listed in Table 4.1. The analysis clearly indicates that the MDCS design yields a broader transmission band and lower insertion loss. The optimized geometric parameters used in the final design are p = 20 mm, w_b =1.2 mm, and l_b = 9 mm.

4.1.3 Analysis of the Proposed FSR

The finalized FSR design, illustrated in Figure 4.5, is achieved by substituting the ground plane of the broadband absorber with a modified double cross-slot (MDCS) bandpass FSS. In this configuration, the bandpass FSS functions as a metallic reflector within the absorption frequency range, while enabling transmission at a higher frequency band. The structure is numerically analyzed, and its simulated reflection and transmission characteristics are presented in Figure 4.6. The results reveal that the proposed FSR supports a transmission band centered around 11.46 GHz and demonstrates broadband absorption covering approximately 3.1–8.7 GHz. In the

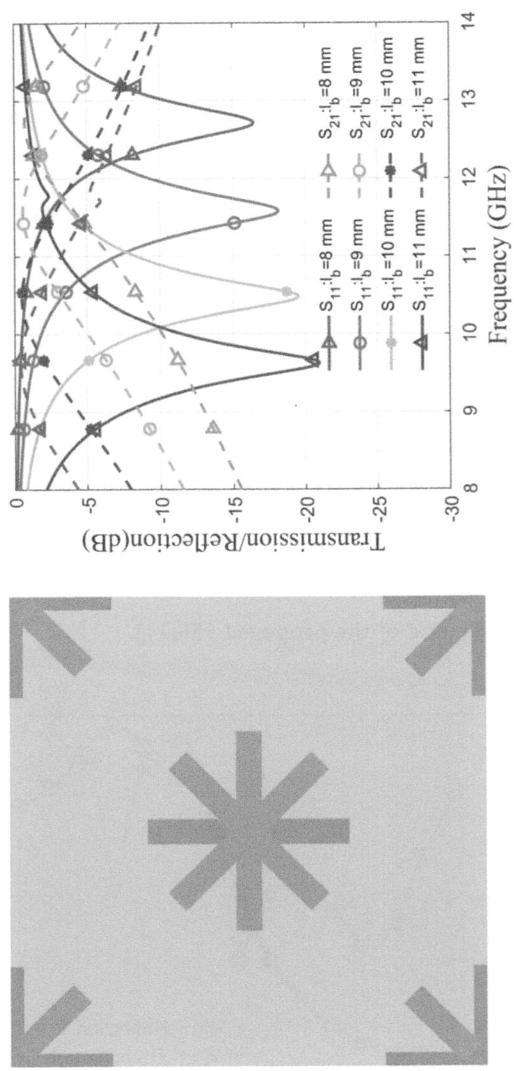

Figure 4.4 MDCS shaped bandpass frequency selective surface: (a) unit cell schematic and (b) simulated response [1].

Table 4.1 Performance Comparison of DCS and MDCS-Type Bandpass FSS with l_b = 9 mm [1]

Structure	*Transmission Frequency f_T*	*Minimum Insertion Loss*	*−10 dB Reflection Bandwidth*
DCS	10.9 GHz	1.82 dB	157 MHz
MDCS	11.56 GHz	0.56 dB	752 MHz

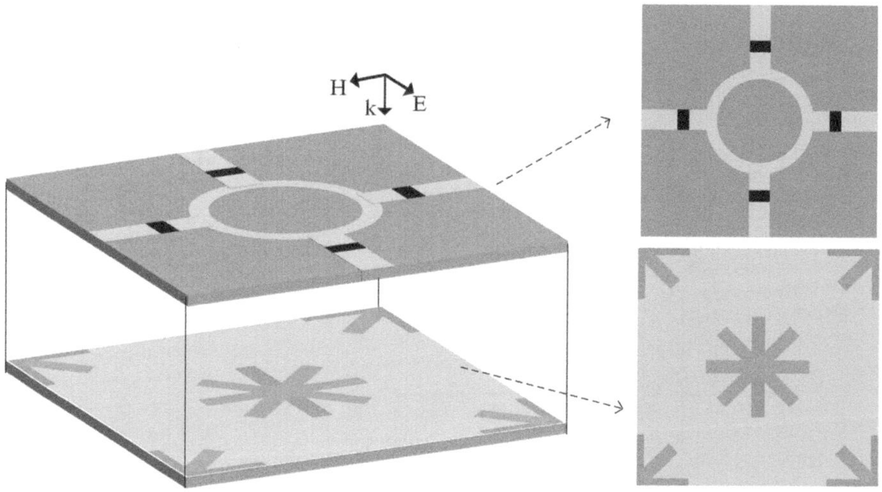

Figure 4.5 Unit cell schematic of the proposed FSR [1].

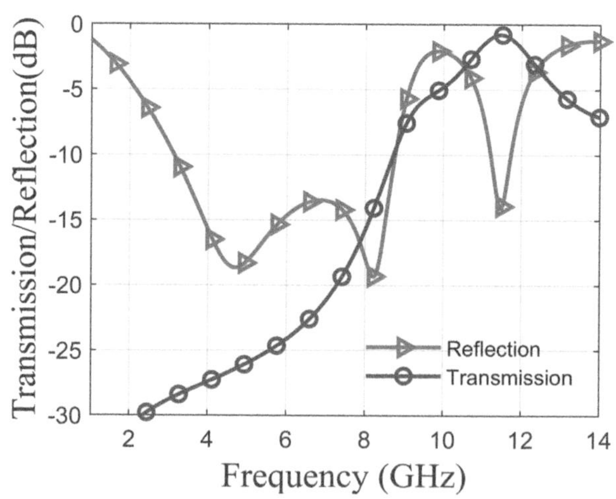

Figure 4.6 Simulated reflection and transmission coefficients of the proposed FSR [1].

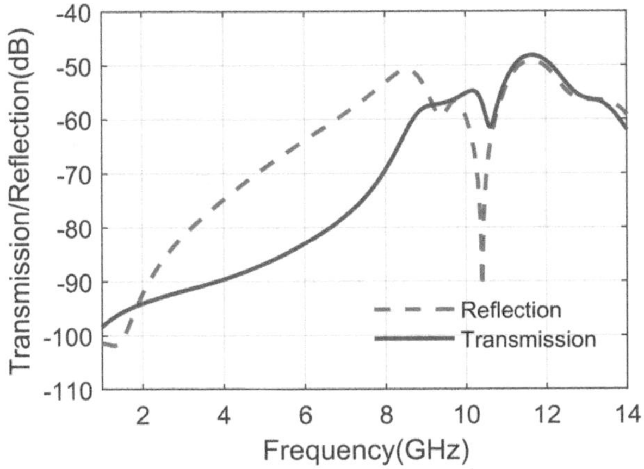

Figure 4.7 Transmission and reflection coefficients of the proposed FSR under cross-polarization [1].

absorption region, both reflection and transmission coefficients fall below −10 dB, indicating effective suppression. The structure also offers low insertion loss, with a minimum of 0.7 dB in the passband, and achieves a −10 dB transmission bandwidth of about 520 MHz, corresponding to a fractional bandwidth of 4.53%.

To further confirm the absorption characteristics, the proposed FSR is also examined under cross-polarized excitation. The simulated reflection and transmission responses for cross-polarized waves are illustrated in Figure 4.7. The observed minimal response under cross-polarization indicates the absence of polarization conversion, thereby reinforcing the validity of the absorption behavior.

Figure 4.8 illustrates the performance of the proposed FSR under different polarization angles of the incident electromagnetic wave. It is evident that the reflection and transmission characteristics remain consistent as the polarization angle varies from 0° to 45°. This stable response confirms that the symmetric design ensures polarization-insensitive behavior.

The angular stability of the designed FSR is assessed through simulated results at different angles of incidence, as illustrated in Figure 4.9. When the incidence angle exceeds 30°, the reflection coefficient rises above the −10 dB threshold, indicating a decline in absorption performance. Therefore, the proposed structure maintains effective operation up to an incident angle of 30°, beyond which its efficiency deteriorates.

4.2 EQUIVALENT CIRCUIT ANALYSIS

To gain insight into the operational mechanism of the proposed FSR, its equivalent circuit model (ECM) is presented in Figure 4.10, where the

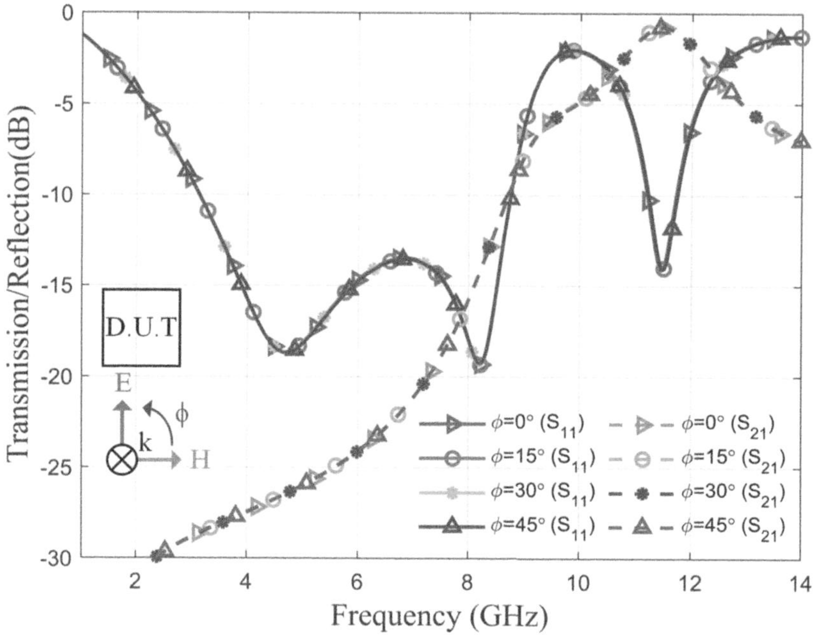

Figure 4.8 Simulated response of the proposed FSR with varying polarization angles [1].

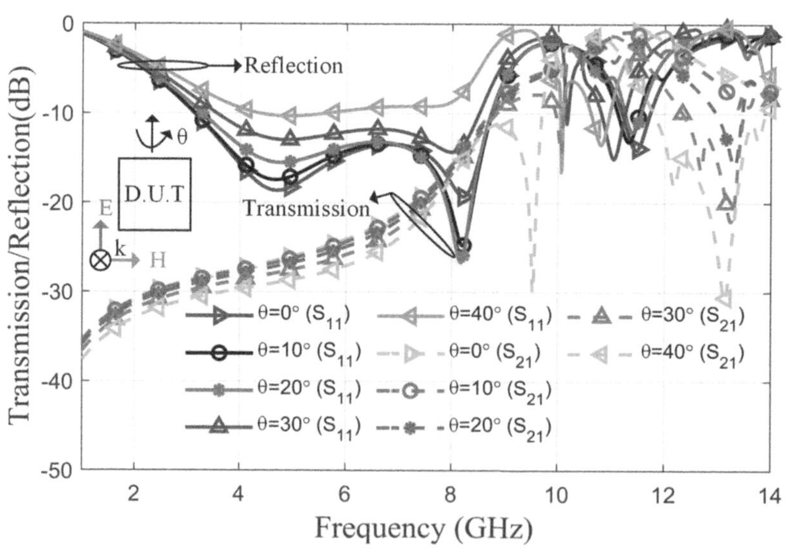

Figure 4.9 Simulated response of the proposed FSR with varying incident angles [1].

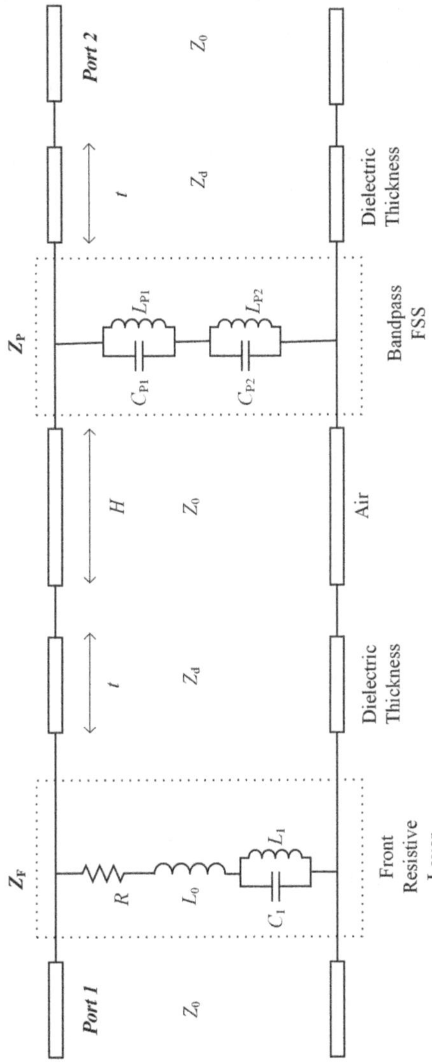

Figure 4.10 Equivalent circuit model of the proposed FSR [1].

individual ECMs of the broadband absorber and the bandpass FSS are connected in series. The top resistive layer is modeled using a parallel LC circuit (L_1 and C_1) representing the central circular resonator, while the outer rectangular strip and integrated resistor are represented by a series combination of R and L_0. In the bottom layer, the center and corner slots of the bandpass FSS are represented by two sets of series LC resonators (L_{P1}–C_{P1} and L_{P2}–C_{P2}). The dielectric substrates of the top and bottom layers are modeled as transmission lines with length t and characteristic impedance $Z_d = Z_0 / \sqrt{\epsilon_r}$, where Z_0 is the impedance of free space. Additionally, the air gap separating the two layers is represented by another transmission line section of length H and impedance Z_0. The overall ABCD matrix of the ECM is derived by sequential multiplication of the individual matrix components as given:

$$\begin{bmatrix} A & B \\ C & D \end{bmatrix} = \begin{bmatrix} 1 & 0 \\ 1/Z_F & 1 \end{bmatrix} \times \begin{bmatrix} \cos\beta t & jZ_d \sin\beta t \\ j\sin\beta t / Z_d & \cos\beta t \end{bmatrix} \times$$
$$\begin{bmatrix} \cos\beta_0 H & jZ_0 \sin\beta_0 H \\ j\sin\beta_0 H / Z_0 & \cos\beta_0 H \end{bmatrix} \times \begin{bmatrix} \cos\beta t & jZ_d \sin\beta t \\ j\sin\beta t / Z_d & \cos\beta t \end{bmatrix} \times \begin{bmatrix} 1 & 0 \\ 1/Z_P & 1 \end{bmatrix} \tag{4.1}$$

Here, β is defined as $\beta = \beta_0 \sqrt{\mu_r \epsilon_r}$, where β_0 denotes the free-space propagation constant. The terms Z_F and Z_P represent the equivalent impedances of the top resistive layer and the bandpass FSS layer, respectively, and are expressed as follows:

$$Z_F = R + j\omega\left(L_0 + \frac{L_1}{1 - \omega^2 L_1 C_1}\right) \tag{4.2}$$

$$Z_P = j\left(\frac{\omega L_{P1}}{1 - \omega^2 L_{P1} C_{P1}} + \frac{\omega L_{P2}}{1 - \omega^2 L_{P2} C_{P2}}\right) \tag{4.3}$$

The reflection coefficient S_{11} and the transmission coefficient S_{21} are determined by computing the S-parameters using the following transformation formula:

$$S_{11} = \frac{A + B/Z_0 - CZ_0 - D}{A + B/Z_0 + CZ_0 + D} \tag{4.4}$$

$$S_{21} = \frac{2}{A + B/Z_0 + CZ_0 + D} \tag{4.5}$$

The absorptivity is calculated:

$$A(\omega) = 1 - |S_{11}|^2 - |S_{21}|^2 \tag{4.6}$$

The equivalent circuit model (ECM) of the designed FSR is analyzed using the Keysight ADS electromagnetic simulation tool. Figures 4.11(a) and 4.11(b) illustrate the simulated ECM results for the broadband absorber and the bandpass FSS, respectively. The broadband absorber's ECM is modeled as a single-port

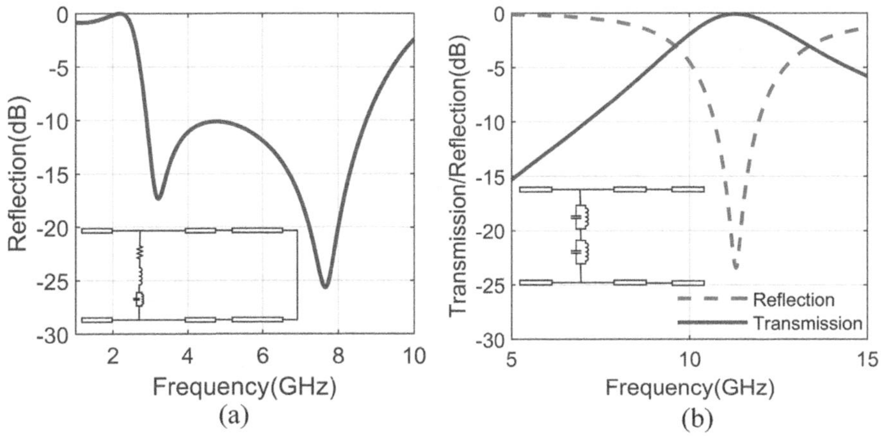

Figure 4.11 Simulated response of equivalent circuit model: (a) broadband absorber and (b) bandpass FSS [1].

network comprising only the front resistive layer, with the second port connected to ground. The optimized parameters for the circuit elements are: $L_0 = 5.14\text{nH}$, $L_1 = 17.4\text{nH}$, $C_1 = 0.302\text{pF}$, $L_{P1} = 0.414\text{nH}$, $C_{P1} = 0.432\text{pF}$, $L_{P2} = 0.414\text{ nH}$, $C_{P2} = 0.432\text{ pF}$, $R = 200\ \Omega$, $Z_d = 179.72\ \Omega$, $Z_0 = 377\ \Omega$. The comparison between the ECM simulation and the CST-based electromagnetic simulation for the complete FSR is shown in Figure 4.12. The ECM demonstrates a strong correlation with CST results, accurately replicating both the absorption range (2.7–8.8 GHz) and the transmission peak at 11.4 GHz.

4.3 EFFECT OF BANDPASS FSS ON THE PERFORMANCE OF FSR

The designed FSR demonstrates a high-frequency transmission band centered around 11.46 GHz, positioned above the lower absorption range spanning 3.1–8.7 GHz. To evaluate the efficiency of this transition from absorption to transmission, a parameter known as the transition rate (TR) is introduced. This metric represents the ratio between the transmission and absorption frequencies and provides a measure of how rapidly the structure shifts from absorbing to transmitting behavior, as defined here:

$$TR = \frac{f_T}{f_a} \tag{4.7}$$

Here, f_T denotes the transmission frequency, while f_a represents the peak absorption frequency. The emergence of the higher transmission band is attributed to the bandpass FSS layer; hence, this band can be adjusted by

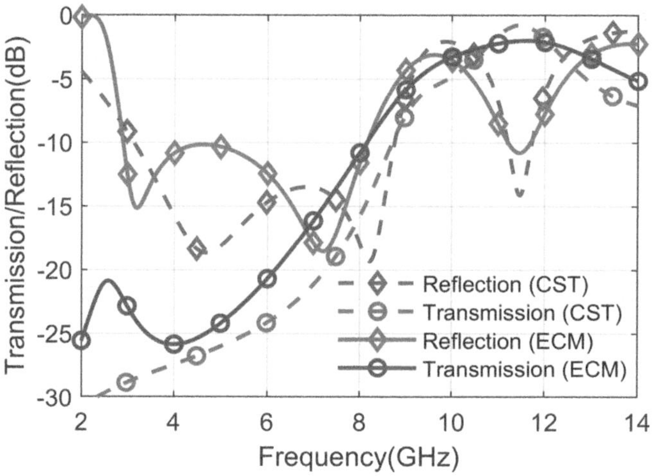

Figure 4.12 Comparison of the reflection and transmission response of the ECM and full wave simulation of the proposed FSR [1].

modifying the design parameters of the FSS. To explore the tunable range of the upper transmission band in the proposed FSR, a parametric study is conducted by varying the slot length of the bandpass FSS. Figure 4.13 illustrates how changes in the slot length l_b influence the transmission frequency. It is observed that increasing l_b results in a downward shift in the transmission frequency. The detailed impact of l_b variation on the performance characteristics of the proposed FSR, as inferred from Figure 4.13, is summarized in Table 4.2.

Based on the data presented in Table 4.2, it is evident that increasing the slot length l_b from 5 mm to 12 mm causes the transmission frequency f_T to shift downward from 18.1 GHz to 9.6 GHz, with the transition rate (TR) varying between 2.03 and 1.39. The insertion loss at f_T initially decreases as l_b increases from 5 mm to 10 mm but begins to rise again once l_b exceeds 10 mm. Moreover, as f_T is reduced to 9.6 GHz, the absorption band also narrows significantly to a range of 3.0–6.9 GHz. Therefore, to ensure an optimal range of f_T with insertion loss kept under 2.0 dB and with minimal change to the lower absorption band, the slot length l_b should ideally lie between 6 mm and 10 mm. The achievable f_T range corresponding to this slot dimension interval is:

$$10.6 \text{ GHz} \leq f_t \leq 16.2 \text{ GHz} \tag{4.8}$$

A *TR* of 1.30 and 1.82 is obtained for transmission frequencies f_T of 10.6 GHz and 16.2 GHz, respectively. This indicates that in the proposed FSR design, the upper transmission frequency can be adjusted within the range of 10.6 GHz to 16.2 GHz by modifying the bandpass FSS slot length l_b from 10 mm down to 6 mm.

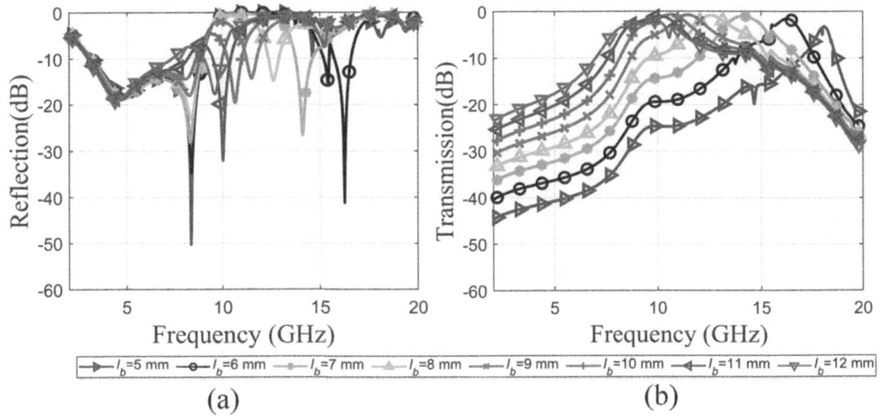

Figure 4.13 Simulated response of the proposed FSR with slot length l_b: (a) reflection and (b) transmission coefficients [1].

Table 4.2 Variation of Performance Parameters with Varying l_b [1]

l_b (mm)	*Transmission Frequency f_T*	*Insertion Loss @ f_T GHz*	*Absorption*	*TR*
5	18.1 GHz	3.25 dB	3.2 – 8.9 GHz	2.03
6	16.2 GHz	1.56 dB	3.2 – 8.9 GHz	1.82
7	14.1 GHz	1.04 dB	3.2 – 8.8 GHz	1.60
8	12.6 GHz	0.85 dB	3.1 – 8.8 GHz	1.43
9	11.4 GHz	0.74 dB	3.1 – 8.7 GHz	1.31
10	10.6 GHz	0.68 dB	3.0 – 8.1 GHz	1.30
11	10.0 GHz	1.04 dB	3.0 – 7.5 GHz	1.33
12	9.6 GHz	2.07 dB	3.0 – 6.9 GHz	1.39

4.4 EXPERIMENTAL VERIFICATION

To experimentally verify the performance of the proposed FSR, a prototype consisting of a 13×14 unit cell array was fabricated for both structural layers. Surface-mounted chip resistors of 200 Ω (model CRCW0603200RFKEA from VISHAY36) were attached to the front side of the top layer. The fabricated structure, having total dimensions of 260 mm × 280 mm, is displayed in Figure 4.14. A plastic spacer standoff was used to maintain the air gap between the two substrate layers, as shown in Figure 4.14(d).

The performance of the fabricated prototype was evaluated using a free-space measurement setup. For transmission and reflection testing across the S, C, J, and X bands, appropriate horn antennas were employed in combination with the Keysight PNA Network Analyzer (model N5224B). The experimental arrangement for each measurement type is illustrated in

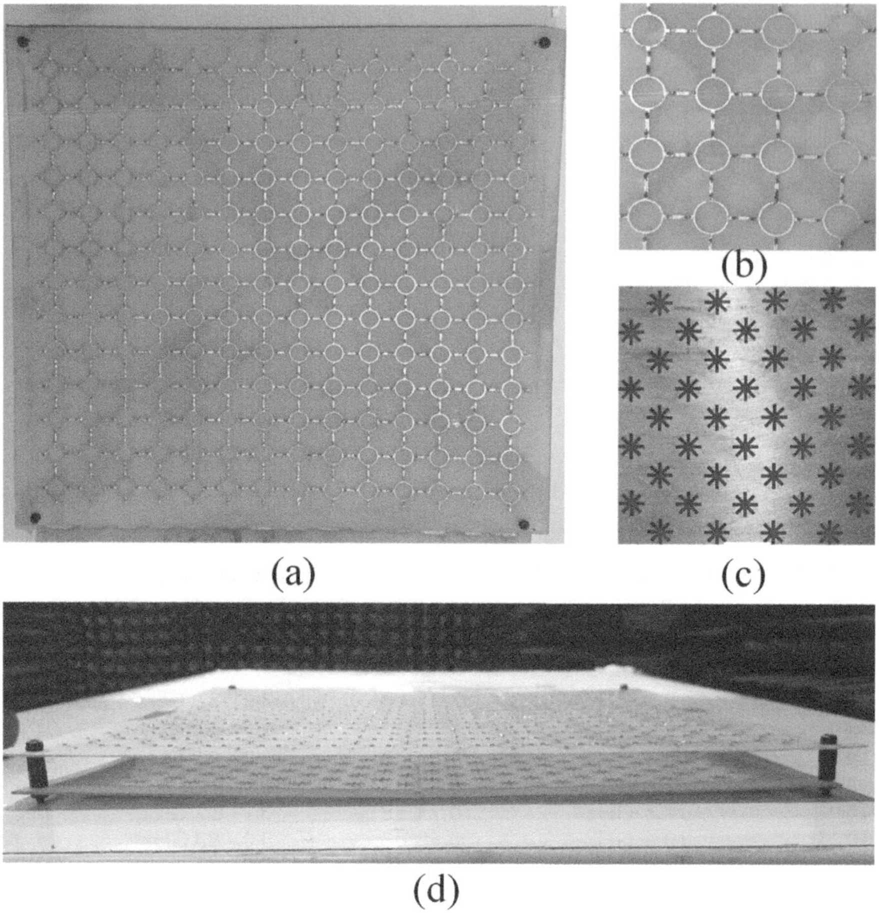

Figure 4.14 Fabricated prototype of the proposed FSR: (a) top layer, (b) enlarged view of top layer, (c) enlarged view of bottom layer, and (d) side view [1].

Figure 4.15. During reflection testing, a perfect electric conductor (PEC) sheet served as the reference, which was then substituted with the fabricated sample to capture the reflection coefficient. For transmission analysis, the reference was obtained by measuring the direct path between the transmitting and receiving horn antennas. The effective transmission coefficient was calculated by comparing the measured data from the direct path with that of the prototype. Figure 4.16 presents the measured transmission and reflection responses alongside the simulated results, showing a good level of consistency between them.

Table 4.3 provides a comparative analysis of the proposed A-T FSR against previously reported designs in references [2–5]. The proposed structure demonstrates a polarization-insensitive behavior and offers a broader

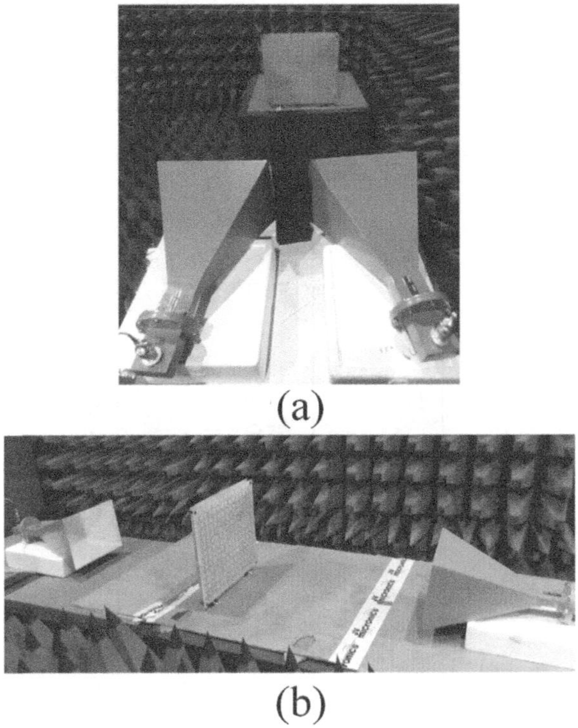

Figure 4.15 Measurement setup: (a) reflection and (b) transmission measurement [1].

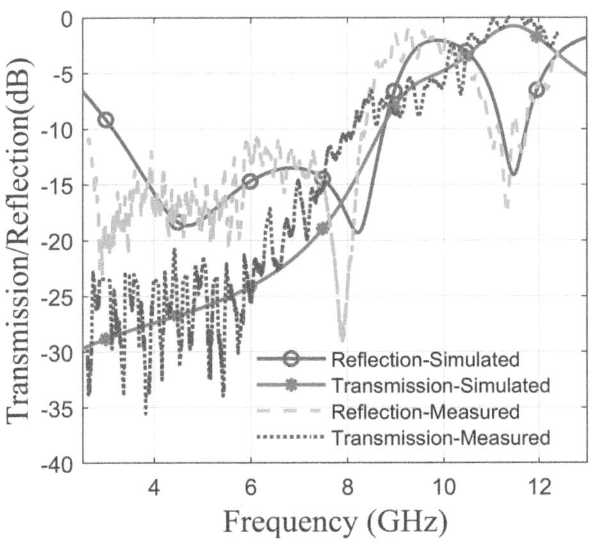

Figure 4.16 Simulated and measured response of the proposed FSR [1].

Table 4.3 Performance Comparison of Proposed A-T FSR [1]

Ref.	f_T /*Mini mum I.L.*	*Absorption*	*No. of Layers*	*Lumped Elements per Unit Cell*	*Size*	*Thickness*	*Polarization*
[2]	5.6 GHz / 0.2 dB	2.8 – 5.0 GHz (56.41%)	2	1	$0.28\lambda_L$	$0.122\lambda_L$	Sensitive
[3]	12.76 GHz/0.65 dB	6.10 – 10.98 GHz (57.14%)	2	2	$0.366\lambda_L \times 0.183\lambda_L$	$0.122\lambda_L$	Sensitive
[4]	10 GHz/0.5 dB	2.4 – 7.1 GHz (98.94%)	4	6	$0.096\lambda_L$	$0.12\lambda_L$	Dual polarized
[5]	6.1 GHz/0.37 dB	1.6 – 4.3 GHz (91.5%)	4	8	$0.16\lambda_L$	$0.158\lambda_L$	Dual polarized
This work	11.4 GHz/0.7 dB	3.1 – 8.7 GHz (94.91%)	2	4	$0.206\lambda_L$	$0.14\lambda_L$	Insensitive

Note: f_T : transmission frequency; λ_L : free space wavelength at lowest frequency of −10 dB reflection; I.L.: the insertion loss.

absorption bandwidth compared to those presented in [2, 3], and [5]. Additionally, the design features a two-layer configuration and utilizes fewer lumped components than the FSRs reported in [4] and [5].

4.5 CONCLUSION

This chapter introduces a frequency selective rasorber (FSR) design that features wideband absorption and a relatively high-frequency transmission band (A-T functionality). The structure consists of two layers: a resistive layer and a bandpass filtering layer. The proposed FSR achieves broad absorption from 3.1 GHz to 8.7 GHz, maintaining over 80% absorptivity, and supports a high transmission band centered at 11.46 GHz with an insertion loss of just 0.7 dB and a –10 dB reflection bandwidth of 520 MHz (ranging from 11.206 to 11.726 GHz). An equivalent circuit model (ECM) is developed to explain the underlying operational mechanism. A detailed parametric analysis is also conducted to identify the tunable range of the high transmission frequency. Finally, a physical prototype is fabricated, and experimental measurements are performed to validate the simulated results.

REFERENCES

1. M. M. Zargar et al., "Polarization-Insensitive Frequency Selective Rasorber With Higher Transmission Band Using Modified Double Cross Slot Based Bandpass Layer," International Journal of RF and Microwave Computer-Aided Engineering, vol. 31, no. 8, e22733, 2021.
2. Z. Wang et al., "A High-Transmittance Frequency-Selective Rasorber Based on Dipole Arrays," IEEE Access, vol. 6, pp. 31367–31374, 2018.
3. M. Guo, Z. Sun, D. Sang, X. Jia and Y. Fu, "Design of Frequency-Selective Rasorbers Based on Centrosymmetric Bended-Strip Resonator," IEEE Access, vol. 7, pp. 24964–24970, 2019.
4. Q. Chen, D. Sang, M. Guo and Y. Fu, "Miniaturized Frequency-Selective Rasorber With a Wide Transmission Band Using Circular Spiral Resonator," IEEE Trans. Antenn. Propag., vol. 67, no. 2, pp. 1045–1052, Feb. 2019.
5. Z. Wang, J. Fu, Q. Zeng, M. Song and T. A. Denidni, "Wideband Transmissive Frequency-Selective Absorber," IEEE Antenn. Wirel. Propag. Lett., vol. 18, no. 7, pp. 1443–1447, Jul. 2019.

Chapter 5

T-A-T FSR

In the preceding chapter, an A-T type FSR featuring broadband absorption and a higher-frequency transmission band was discussed. This chapter focuses on the development of a T-A-T-type FSR that supports dual-band transmission. The design integrates a dual-band bandpass frequency selective surface (FSS) with a broadband absorber, resulting in a structure that provides wideband absorption between two distinct transmission windows. The proposed FSR is a compact two-layer configuration and is well suited for radar cross-section (RCS) reduction in dual-band antenna systems. The operational behavior of the structure is examined through an equivalent circuit model (ECM). A prototype is fabricated, and experimental measurements are performed, showing strong correlation with the simulation results, thereby validating the design's effectiveness.

5.1 DESIGN AND ANALYSIS OF THE FSR

An FSR is typically realized using a two-layer structure comprising a lossy layer and a lossless layer. The lossy layer, which incorporates components such as lumped resistors, is responsible for achieving broadband absorption. In contrast, the lossless layer, functioning as a bandpass FSS, facilitates the desired transmission bands. The design process begins with the development of a dual-band bandpass FSS, followed by the design of a broadband absorber whose unit cell dimensions are aligned with those of the bandpass FSS. Additionally, the absorption band is tailored to lie between the two transmission bands. For this to work effectively, the bandpass FSS must behave as a reflective surface within the absorption band. The final FSR configuration is obtained by substituting the ground plane of the absorber with the designed bandpass FSS.

5.1.1 Bandpass FSS

The schematic of the bandpass FSS unit cell is illustrated in Figure 5.1(a), featuring a Jerusalem cross-shaped slot etched into a metal-coated substrate.

DOI: 10.1201/9781003644323-5

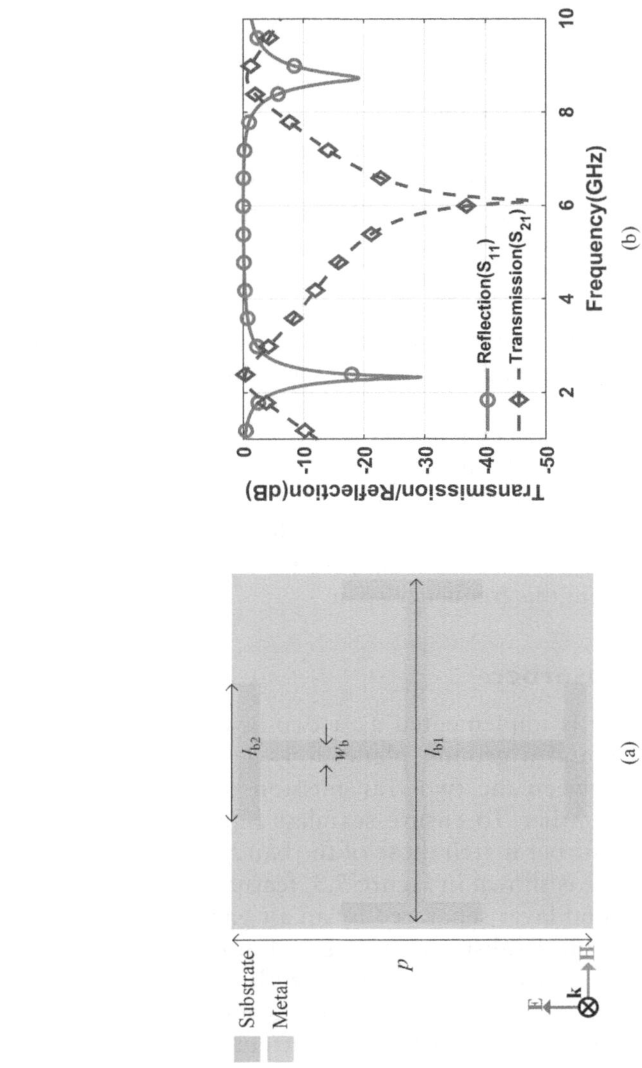

Figure 5.1 Jerusalem-shaped bandpass FSS: (a) unit cell schematic and (b) simulated reflection and transmission coefficients [1].

The structure utilizes an FR-4 substrate with a thickness of 0.8 mm, a dielectric constant (ϵ_r) of 4.4, a loss tangent (tan δ) of 0.02, and a metal layer thickness of 0.035 mm. The Jerusalem cross geometry is formed by introducing a rectangular slot of length l_{b2} at each arm end of a primary cross-slot of length l_{b1}. Electromagnetic simulation of the unit cell is carried out using CST. The optimized geometric parameters are $p = 25$ mm, $l_{b1} = 24$ mm, $l_{b2} = 12$ mm, $w_b = 1.3$ mm.

The simulated transmission and reflection coefficients of the bandpass FSS are shown in Figure 5.1(b), revealing two distinct transmission bands centered around 2.3 GHz and 8.7 GHz. The rectangular slots of length l_{b2}, positioned at the ends of the cross-slot, are responsible for the higher frequency transmission, whereas the central cross-slot of length l_{b1} governs the lower frequency band. Figures 5.2(a) and 5.2(b) illustrate how the transmission characteristics vary with changes in l_{b1} and l_{b2}, respectively. From Figure 5.2(a), it is evident that increasing $\mathrm{l_{b1}}$ from 20 mm to 24 mm causes a downward shift in the lower transmission frequency (f_{tl}) from 3.1 GHz to 2.3 GHz, while the upper transmission frequency (f_{tu}) remains mostly unchanged. In contrast, Figure 5.2(b) shows that increasing l_{b2} from 10 mm to 14 mm results in a decrease in f_{tu} from 10.0 GHz to 7.7 GHz, with negligible impact on the lower frequency band.

Figure 5.1(b) shows that in the intermediate frequency range between the two transmission bands, specifically from 3.8 GHz to 7.5 GHz, the transmission coefficient drops below –10 dB, while the reflection coefficient approaches 0 dB. This indicates that the bandpass FSS functions effectively as a reflective surface within this frequency span.

5.1.2 Broadband Absorber

The broadband absorber is implemented as a two-layer structure comprising a resistive layer and a ground plane. It is designed so that its absorption band lies between the two transmission frequencies of the bandpass FSS discussed earlier. To ensure seamless integration, the unit cell dimensions of the absorber match those of the bandpass FSS. The unit cell layout of the absorber is shown in Figure 5.3, featuring a resistive top layer and a metallic ground layer separated by an air gap of height H, all built on a 0.8 mm thick FR-4 substrate (with $\varepsilon_r = 4.4$, tanδ = 0.02). The resistive layer consists of a metallic square loop with length l and width w, where lumped resistors of resistance R are placed across the gaps (gap length = g) in the loop. Simulated reflection coefficients for variations in parameters H, l, and R are shown in Figure 5.4. The optimal design values are $R = 200\,\Omega$, $p = 25$, $l = 15$, $w = 0.5$, $t = 0.8$, and $H = 12$ (all dimensions in mm). With these values, the structure achieves broad absorption from 3.3 GHz to 7.5 GHz, maintaining a reflection coefficient below –10 dB across this range.

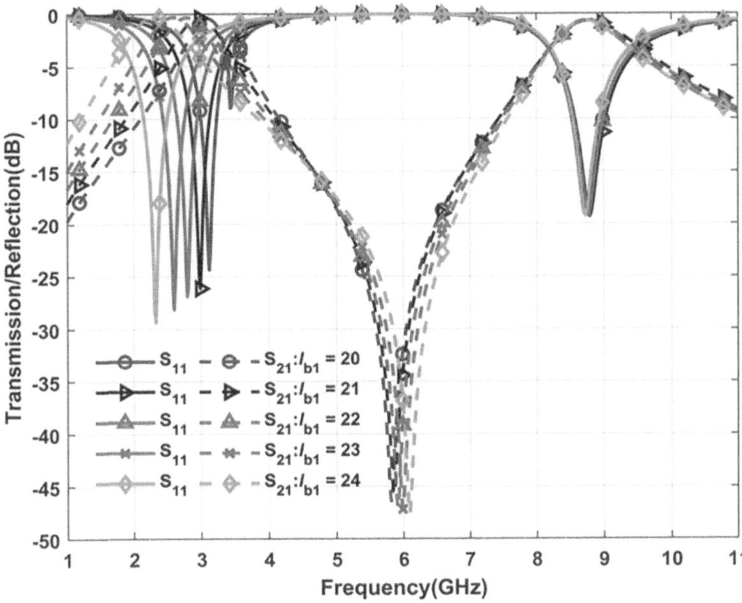

(a)

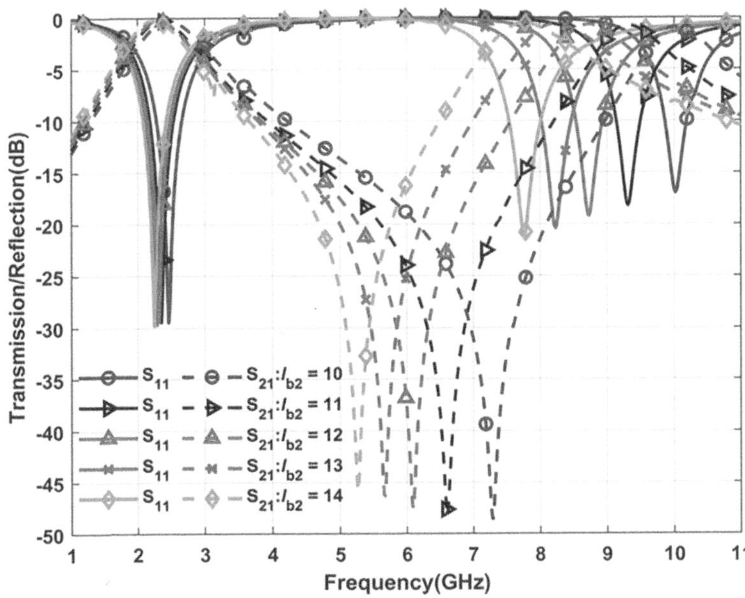

(b)

Figure 5.2 Simulated reflection and transmission coefficients of bandpass FSS with varying (a) $\boldsymbol{l}_{b1}$ (in mm) and (b) l_{b2} (in mm) [1].

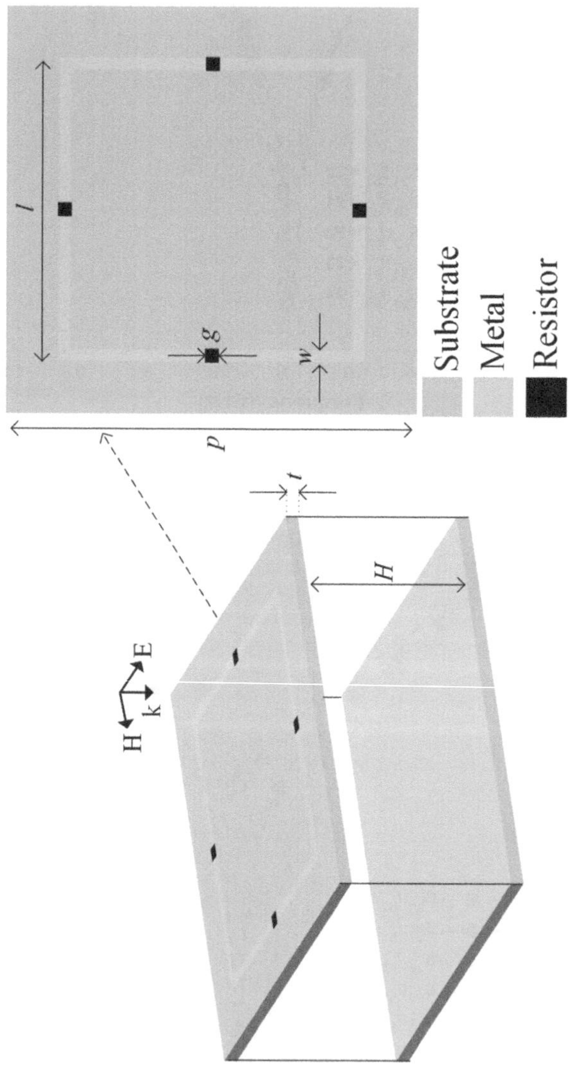

Figure 5.3 Unit cell schematic of the broadband absorber [1].

5.1.3 Proposed FSR

As illustrated in Figure 5.5, the proposed FSR is developed by substituting the ground plane of the broadband absorber with a Jerusalem cross-shaped bandpass frequency selective surface (FSS). The performance of the resulting structure is analyzed, and the simulated transmission and reflection coefficients are shown in Figure 5.6. The FSR demonstrates two distinct transmission bands centered at 2.1 GHz and 8.8 GHz. At the lower frequency of 2.1 GHz, an insertion loss of 1.4 dB is recorded within a –10 dB reflection bandwidth of approximately 443 MHz (1.972–2.415 GHz). For the higher frequency at 8.8 GHz, the structure achieves an insertion loss of 1.5 dB within a –10 dB reflection bandwidth of about 496 MHz (8.579–9.075 GHz). Between these two passbands, the structure exhibits wideband absorption from 3.6 GHz to 7.4 GHz, covering 69.09% of the frequency span, where both the transmission and reflection levels remain below –10 dB. The absorption characteristics of the proposed FSR are shown in Figure 5.7.

The performance of the proposed FSR is evaluated for various polarization angles of the incident wave, as illustrated in Figure 5.8. The reflection and transmission characteristics remain consistent for polarization angles ranging from 0° to 45°, indicating that the proposed FSR maintains polarization-insensitive behavior.

The performance of the proposed FSR is further evaluated under oblique incidence for a TE-polarized wave. Figure 5.9 presents the reflection and transmission responses for incident angles ranging from 0° to 40°. As the angle of incidence increases, the insertion loss at both transmission frequencies also rises. However, the increase in loss is more prominent at the higher transmission frequency compared to the lower one. Notably, the insertion loss at the upper transmission band stays within 3 dB up to an incident angle of 40°, beyond which it exceeds this threshold. Therefore, the proposed FSR demonstrates satisfactory performance for oblique incidence up to 40°.

5.2 EQUIVALENT CIRCUIT MODEL

The operational mechanism of the proposed FSR can be understood through the analysis of its equivalent circuit model (ECM). As illustrated in Figure 5.10, the ECM represents a two-port network comprising the circuit equivalents of both the resistive and bandpass FSS layers. The resistive square loop located on the front layer is modeled using a series R–L_1–C_1 configuration.

The dual-band bandpass layer located at the rear side is represented by a series combination of two parallel LC resonators: $L_{P1} - C_{P1}$ and $L_{P2} - C_{P2}$. The air spacer separating the two layers is modeled as a transmission line with a characteristic impedance equal to the intrinsic impedance of free space (Z_0). Similarly, the dielectric substrate is modeled as a transmission

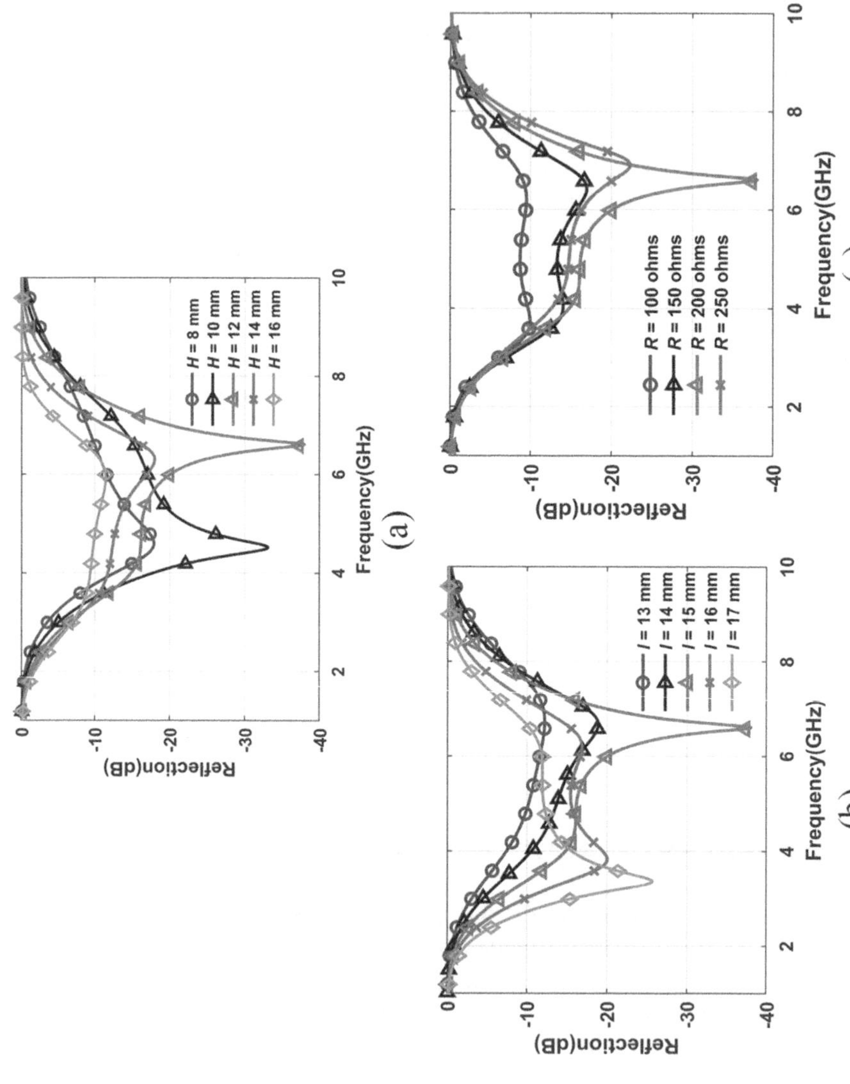

Figure 5.4 Reflection coefficient of broadband absorber with varying (a) **H**, (b) l, and (c) **R** [1].

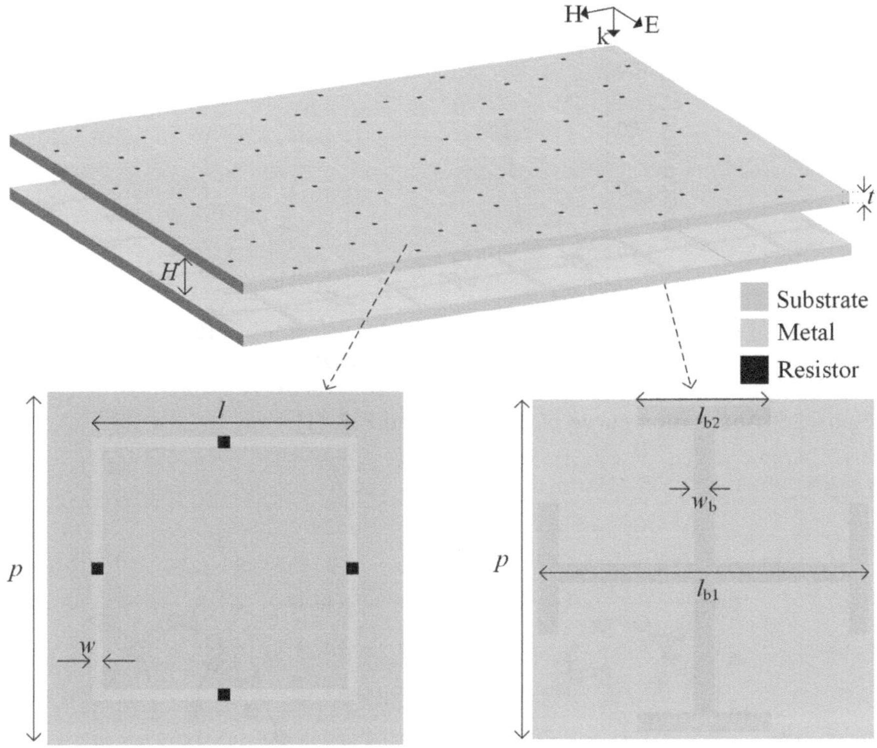

Figure 5.5 Schematic of the proposed FSR [1].

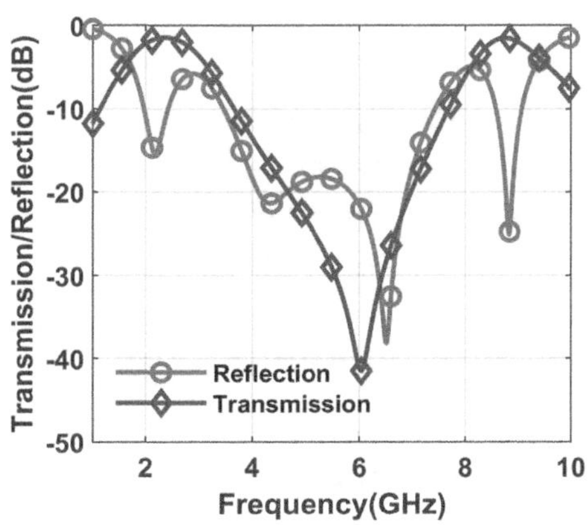

Figure 5.6 Simulated reflection and transmission coefficients of the proposed FSR [1].

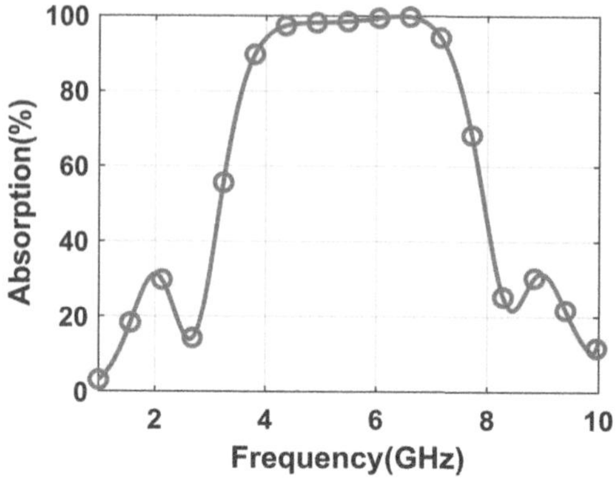

Figure 5.7 Absorption curve of the proposed FSR [1].

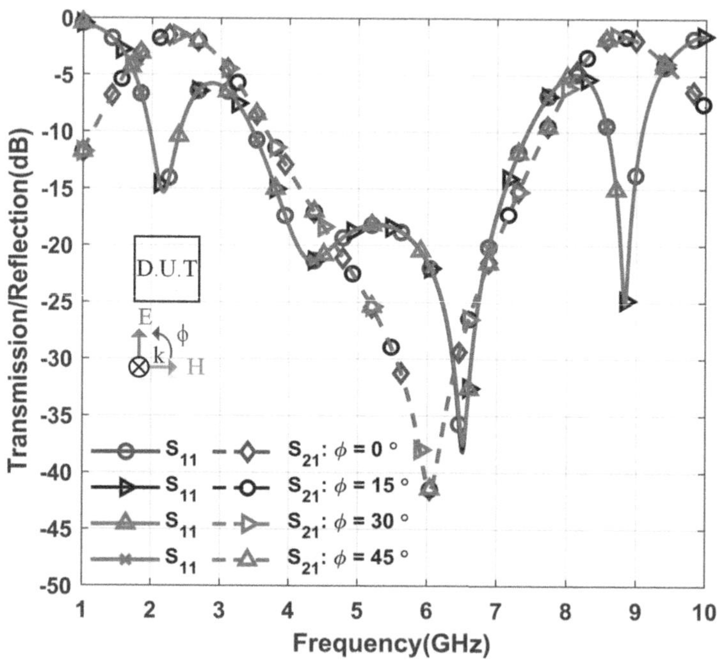

Figure 5.8 Simulated reflection and transmission coefficients of the proposed FSR under various polarization angles of the incident wave [1].

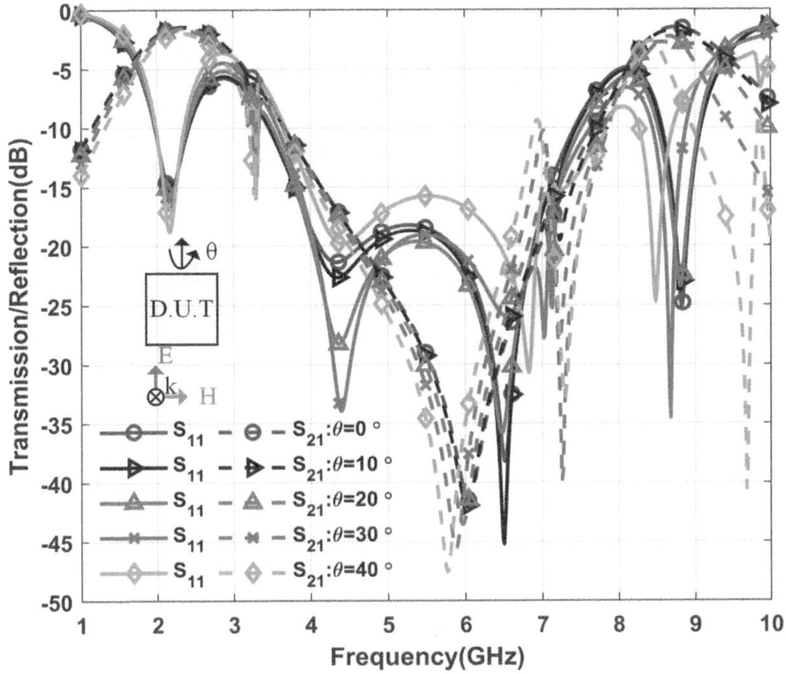

Figure 5.9 Simulated reflection and transmission coefficients of the proposed FSR under different incident angles for a TE polarized wave [1].

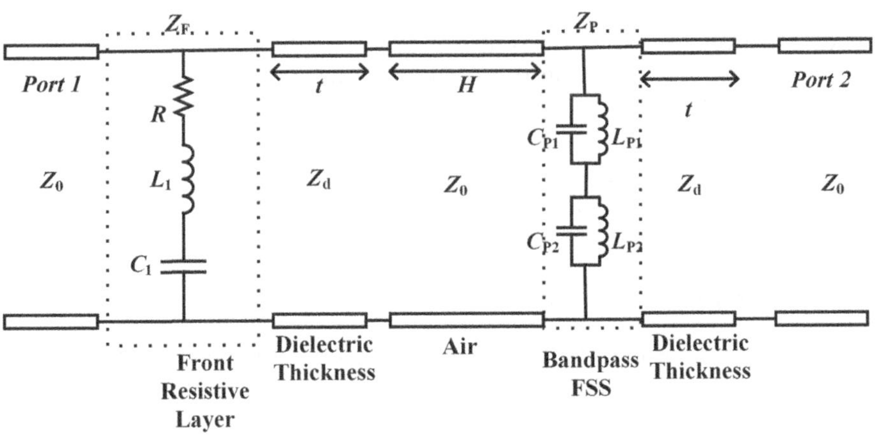

Figure 5.10 ECM of the proposed FSR [1].

line segment with length equal to the substrate thickness t and characteristic impedance Z_d, where $Z_d = Z_0 / \sqrt{\epsilon_r}$. Both the input and output ports of the ECM are terminated with free-space impedance. The complete ABCD matrix for the equivalent circuit is derived by sequentially multiplying the ABCD matrices of the individual components:

$$\begin{bmatrix} A & B \\ C & D \end{bmatrix} = \begin{bmatrix} 1 & 0 \\ 1/Z_F & 1 \end{bmatrix} \times \begin{bmatrix} \cos\beta t & jZ_d \sin\beta t \\ j\sin\beta t / Z_d & \cos\beta t \end{bmatrix} \times \begin{bmatrix} \cos\beta_0 H & jZ_0 \sin\beta_0 H \\ j\sin\beta_0 H / Z_0 & \cos\beta_0 H \end{bmatrix} \times \begin{bmatrix} \cos\beta t & jZ_d \sin\beta t \\ j\sin\beta t / Z_d & \cos\beta t \end{bmatrix} \times \begin{bmatrix} 1 & 0 \\ 1/Z_P & 1 \end{bmatrix} \tag{5.1}$$

In Equation (5.1), the parameter β denotes the propagation constant within the dielectric substrate and is expressed as $\beta = \beta_0 \sqrt{\mu_r \epsilon_r}$, where μ_r and ε_r represent the substrate's relative permeability and permittivity, respectively, and β_0 is the propagation constant in free space. The terms Z_F and Z_P correspond to the equivalent impedances of the resistive layer and the bandpass FSS layer, respectively. Their analytical expressions are provided in Equations (5.2) and (5.3).

$$Z_F = R + j\left(\omega L_1 - \frac{1}{\omega C_1}\right) \tag{5.2}$$

$$Z_P = j\left(\frac{\omega L_{P1}}{1 - \omega^2 L_{P1} C_{P1}} + \frac{\omega L_{P2}}{1 - \omega^2 L_{P2} C_{P2}}\right) \tag{5.3}$$

The reflection (S_{11}) and transmission (S_{21}) coefficients are derived from the ABCD parameters using the standard conversion formulas outlined in Equations (5.4) and (5.5).

$$S_{11} = \frac{A + B/Z_0 - CZ_0 - D}{A + B/Z_0 + CZ_0 + D} \tag{5.4}$$

$$S_{21} = \frac{2}{A + B/Z_0 + CZ_0 + D} \tag{5.5}$$

The absorptivity is determined based on the S-parameters using the expression provided in Equation (5.6).

$$A(\omega) = 1 - |S_{11}|^2 - |S_{21}|^2 \tag{5.6}$$

The equivalent circuit model (ECM) of the proposed FSR is evaluated for its S-parameters using the Keysight ADS simulation tool. The optimized component values used in the ECM are $R = 200\ \Omega, L_1 = 8.96\ \text{nH}, C_1 = 0.11\ \text{pF}$, $L_{P1} = 0.69\ \text{nH}, C_{P1} = 0.49\ \text{pF}, L_{P2} = 7.15\ \text{nH}, C_{P2} = 0.62\ \text{nH}, Z_0 = 377\ \Omega$, $Z_d = 179.72\ \Omega$. Figure 5.11 presents the reflection and transmission

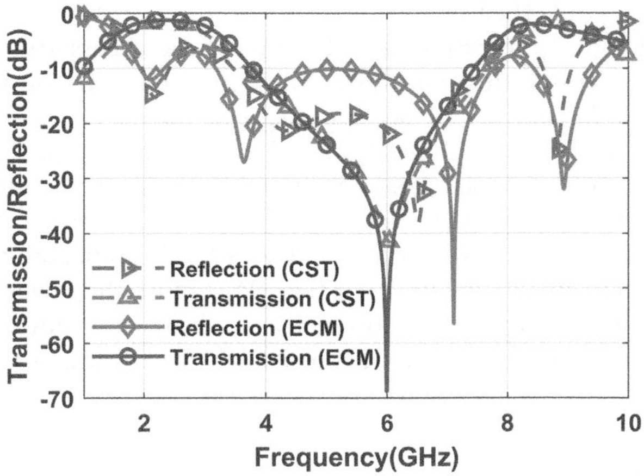

Figure 5.11 Comparison of the reflection and transmission coefficients for the proposed FSR and its corresponding ECM [1].

coefficients obtained from the ECM simulation, along with the CST full-wave simulation results of the proposed FSR. The ECM response exhibits two transmission bands centered at 2.1 GHz and 8.9 GHz, and an intermediate absorption band spanning 3.1 GHz to 7.4 GHz, showing strong correlation with the full-wave simulated performance.

5.3 EFFECT OF PARAMETRIC VARIATIONS ON THE PERFORMANCE OF PROPOSED FSR

The designed FSR demonstrates dual-transmission behavior, with a wide absorption band situated between the two transmission frequencies. The relative positioning of the transmission and absorption bands can be adjusted by varying the parameters of the bandpass FSS. To investigate how these parameter changes affect the separation between the bands, transition ratios are introduced and defined in Equations (5.7) and (5.8).

$$TR_l = \frac{f_{tl}}{f_a} \tag{5.7}$$

$$TR_u = \frac{f_{tu}}{f_a} \tag{5.8}$$

In these equations, TR_l and TR_u represent the transition ratios corresponding to the lower and upper transmission bands, respectively. Here, f_{tl} and f_{tu} denote the center frequencies of the lower and upper transmission bands, while f_a refers to the center frequency of the absorption band.

Parametric analysis is performed on the proposed FSR to evaluate the variation range of the transition ratios. The lower transmission band is primarily influenced by the cross-slot (length l_{b1}) in the Jerusalem cross-shaped bandpass FSS, while the upper transmission band is governed by the rectangular slots (length l_{b2}) positioned at the ends of each cross arm. Accordingly, two separate parametric studies are conducted, focusing on the effects of varying l_{b1} and l_{b2} on the FSR performance.

Figure 5.12 illustrates the reflection and transmission responses of the FSR as the parameter l_{b1} is varied from 25 mm to 20 mm, keeping all other design parameters constant. To aid in understanding, Table 5.1 summarizes the performance metrics of the FSR for each value of l_{b1}. From the table, it is evident that when l_{b1} is equal to the unit cell size (25.0 mm), the lower transmission frequency f_{tl} does not manifest. However, as l_{b1} decreases from 24.9 mm to 20 mm, f_{tl} shifts upward from 1.71 GHz to 2.85 GHz. During this variation, the

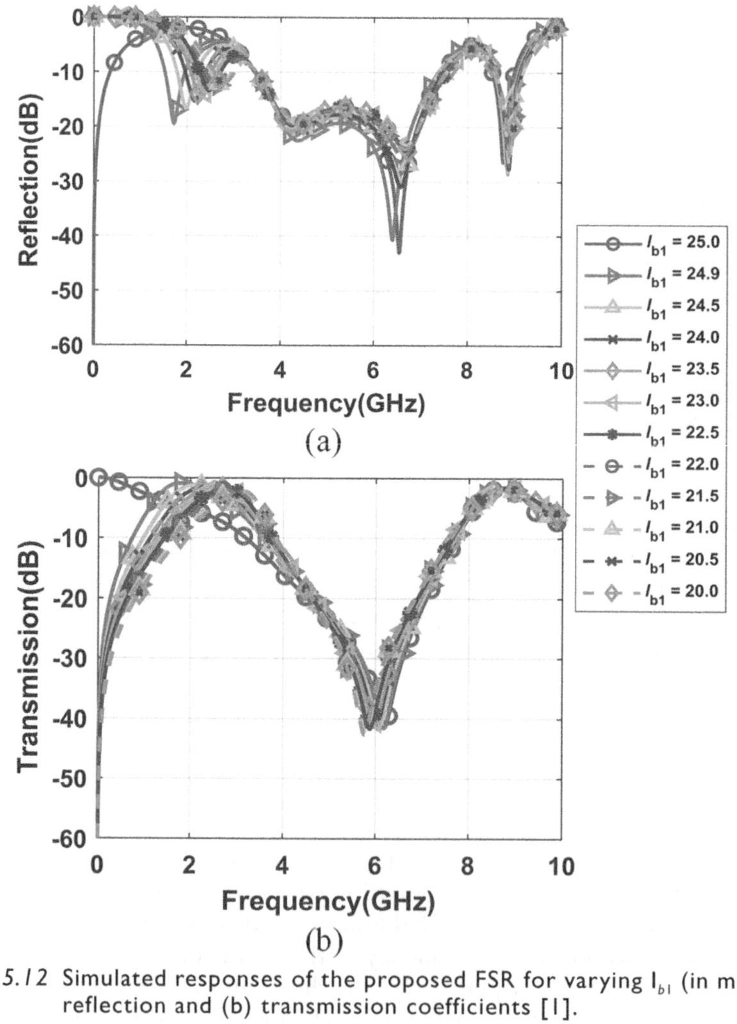

Figure 5.12 Simulated responses of the proposed FSR for varying l_{b1} (in mm): (a) reflection and (b) transmission coefficients [1].

absorption band (with absorptivity > 80%) and the upper transmission frequency f_{tu} (with insertion loss < 2 dB) remain relatively stable, exhibiting only minor fluctuations. It is also observed that, as f_{tl} increases, the corresponding insertion loss also rises, and when f_{tl} exceeds 2.61 GHz (which occurs at l_{b1} = 21.5 mm), the insertion loss surpasses 2 dB. Hence, to maintain insertion loss below 2 dB, the value of f_{tl} can be tuned within the range of 1.71 to 2.61 GHz by adjusting lb_1 from 24.9 mm to 21.5 mm. The corresponding lower and upper bounds of f_{tl} relative to the center frequency of the absorption band (f_a) are expressed in Equations (5.9) and (5.10).

$$f_{tl(\min)} = 1.71GHz = 0.31f_a \tag{5.9}$$

$$f_{tl(\max)} = 2.61GHz = 0.46f_a \tag{5.10}$$

Additionally, Figure 5.13 presents the reflection and transmission characteristics of the FSR as the parameter l_{b2} is varied from 14 mm to 8.5 mm, while keeping all other design parameters unchanged. The corresponding performance data for each value of l_{b2} is provided in Table 5.2. As l_{b2} decreases from 14 mm to 9.0 mm, the upper transmission frequency f_{tu} shifts upward from 8.05 GHz to 10.65 GHz. Throughout this variation, both the absorption band and the lower transmission frequency f_{tl} remain relatively stable with only slight deviations. However, beyond f_{tu} = 10.65 GHz, the reflection coefficient rises above −10 dB, leading to a notable reduction in transmission. To ensure the insertion loss stays below 2 dB, the upper transmission frequency f_{tu} can be tuned between 8.62 GHz and 10.65 GHz by adjusting l_{b2} from 12.5 mm to 9.0 mm, respectively.

Table 5.1 Performance Parameters of the FSR with Varying l_{b1} [1]

l_{b1} (mm)	f_{tl} (GHz)	Minimum I.L in f_{tl} band	Absorption/(f_a) (GHz)	f_{tu} (GHz)	Minimum I.L in f_{tu} band	T_{Rl}	T_{Ru}
25.0	Nil	Nil	3.43–7.45/(5.44)	8.74	1.67 dB	–	1.60
24.9	1.71	0.93 dB	3.56–7.36/(5.46)	8.79	1.50 dB	0.31	1.60
24.5	1.98	1.25 dB	3.61–7.49/(5.55)	8.81	1.66 dB	0.35	1.58
24.0	2.16	1.45 dB	3.67–7.47/(5.57)	8.85	1.57 dB	0.38	1.58
23.5	2.29	1.61 dB	3.68–7.55/(5.61)	8.85	1.76 dB	0.40	1.57
23.0	2.38	1.70 dB	3.72–7.55/(5.63)	8.85	1.76 dB	0.42	1.57
22.5	2.47	1.78 dB	3.74–7.54/(5.64)	8.87	1.73 dB	0.43	1.57
22.0	2.54	1.86 dB	3.75–7.54/(5.64)	8.88	1.72 dB	0.45	1.57
21.5	2.61	1.93 dB	3.76–7.50/(5.63)	8.86	1.66 dB	0.46	1.57
21.0	2.71	2.06 dB	3.83–7.54/(5.68)	8.88	1.74 dB	0.47	1.56
20.5	2.79	2.15 dB	3.84–7.53/(5.68)	8.89	1.69 dB	0.49	1.56
20.0	2.85	2.24 dB	3.86–7.53/(5.69)	8.90	1.66 dB	0.50	1.56

Note: f_{tl} is the center frequency of the lower transmission band; f_a is the center frequency of the broad absorption band; f_{tu} is the center frequency of the upper transmission band; I.L. is the insertion loss at the transmission band.

The lower and upper bounds of f_{tu} relative to the corresponding center frequency of the absorption band (f_a) are specified in Equations (5.11) and (5.12).

$$f_{tu(\min)} = 8.62 \text{ GHz} = 1.56 f_a \tag{5.11}$$

$$f_{tu(\max)} = 10.65 \text{ GHz} = 1.86 f_a \tag{5.12}$$

The ideal range for the lower and upper transmission frequencies of the suggested FSR, relative to the absorption frequency, is specified in Equations (5.13) and (5.14).

$$f_{tl} = 0.31 f_a \text{ to } 0.46 f_a \tag{5.13}$$

$$f_{tu} = 1.56 f_a \text{ to } 1.86 f_a \tag{5.14}$$

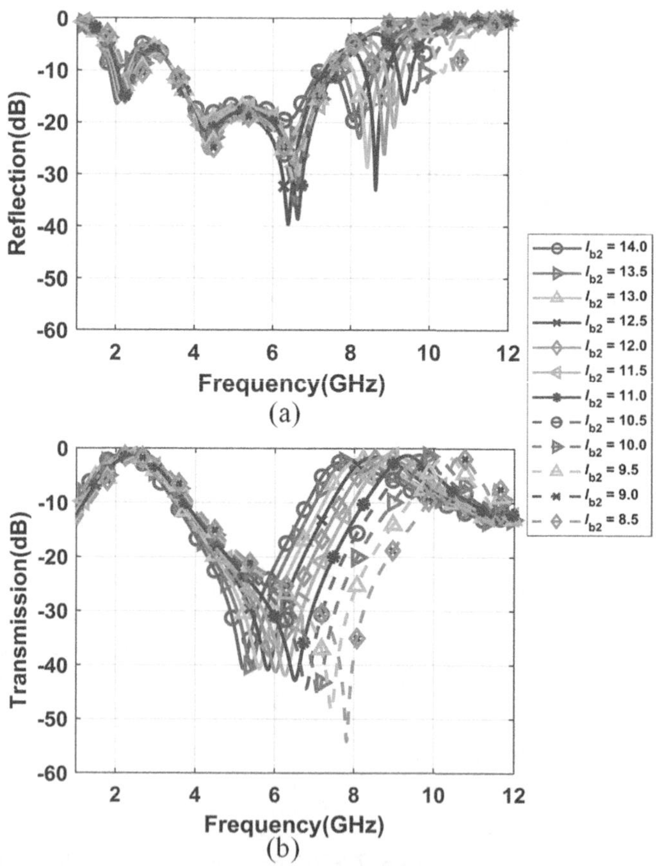

Figure 5.13 Simulated response of the proposed FSR for varying l_{b2} (in mm): (a) reflection, and (b) transmission coefficients [1].

Table 5.2 Performance Parameters of the FSR with Varying lb_2 [1]

l_{b2} (mm)	f_{tl} (GHz)	Minimum I.L. in f_{tl} Band	Absorption/(f_a) (GHz)	f_{tu} (GHz)	Minimum I.L. in f_{tu} Band	T_{Rl}	T_{Ru}
14.0	2.02	1.30	3.47–6.82/(5.14)	8.05	2.61	0.39	1.56
13.5	2.06	1.34	3.54–7.01/(5.46)	8.20	2.20	0.37	1.50
13.0	2.09	1.38	3.59–7.22/(5.27)	8.41	2.06	0.39	1.59
12.5	2.14	1.40	3.67–7.33/(5.50)	8.62	1.57	0.38	1.56
12.0	2.16	1.45	3.67–7.47/(5.57)	8.85	1.57	0.38	1.58
11.5	2.19	1.52	3.68–7.57/(5.62)	9.09	1.48	0.38	1.61
11.0	2.23	1.53	3.72–7.56/(5.64)	9.36	1.33	0.39	1.65
10.5	2.29	1.57	3.78–7.51/(5.64)	9.62	1.22	0.40	1.70
10.0	2.35	1.62	3.83–7.55/(5.69)	9.95	1.27	0.41	1.74
9.5	2.40	1.67	3.88–7.55/(5.71)	10.35	1.63	0.42	1.81
9.0	2.45	1.72	3.92-7.53/(5.72)	10.65	1.52	0.42	1.86
8.5	2.47	1.78	3.94-7.65/(5.79)	–	–	0.42	–

Note: f_{tl} is the center frequency of the lower transmission band; f_a is the center frequency of the broad absorption band; f_{tu} is the center frequency of the upper transmission band; I.L. is the insertion loss at the transmission band.

5.4 EXPERIMENTAL VERIFICATION AND DISCUSSION

A prototype of the FSR, featuring a 10×10 array of unit cells for both resistive and bandpass layers, has been fabricated. The prototype, with dimensions of 250 mm × 250 mm, is depicted in Figure 5.14. A 12 mm air gap between the layers is maintained using plastic spacers, as shown in Figure 5.14(a). In the resistive layer, 200 Ω chip resistors (VISHAY CRC-W0603200RFKEA) are placed within the gaps of a metallic square ring, as illustrated in Figure 5.14(b). The fabricated prototype is tested using a free-space measurement method, employing two standard gain horn antennas connected to a Keysight Vector Network Analyzer (N5224B). The experimental setup for measuring transmission and reflection is shown in Figure 5.15. For reflection measurements, a PEC sheet serves as the reference, which is then substituted with the FSR prototype to calculate the effective reflection coefficient. For transmission measurements, the reference involves direct measurement between the horn antennas. The measured transmission and reflection coefficients, compared with simulated results, are presented in Figure 5.16, demonstrating close alignment and confirming the performance of the proposed FSR.

A comparative analysis of the proposed T-A-T FSR with previously reported designs is summarized in Table 5.3. Unlike the referenced FSRs [2–6], the proposed structure is compact, offers dual transmission functionality, and demonstrates polarization-insensitive behavior.

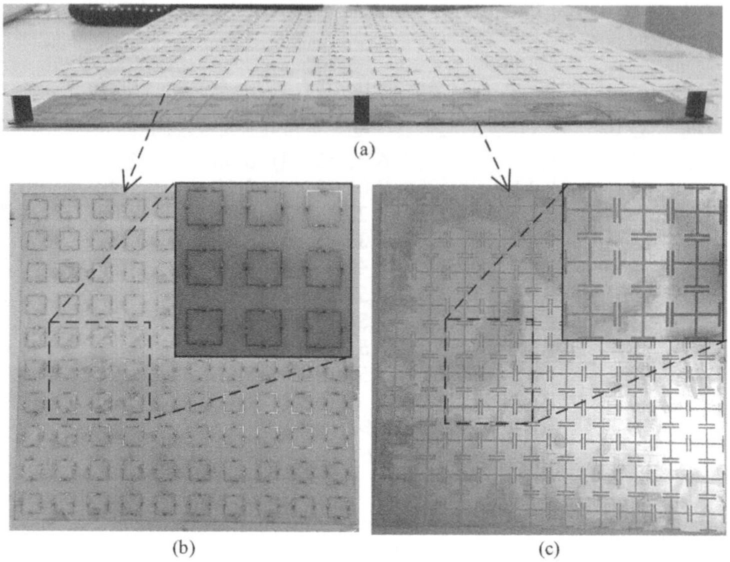

Figure 5.14 Photograph of the fabricated prototype: (a) side view, (b) top, and (c) bottom layers [1].

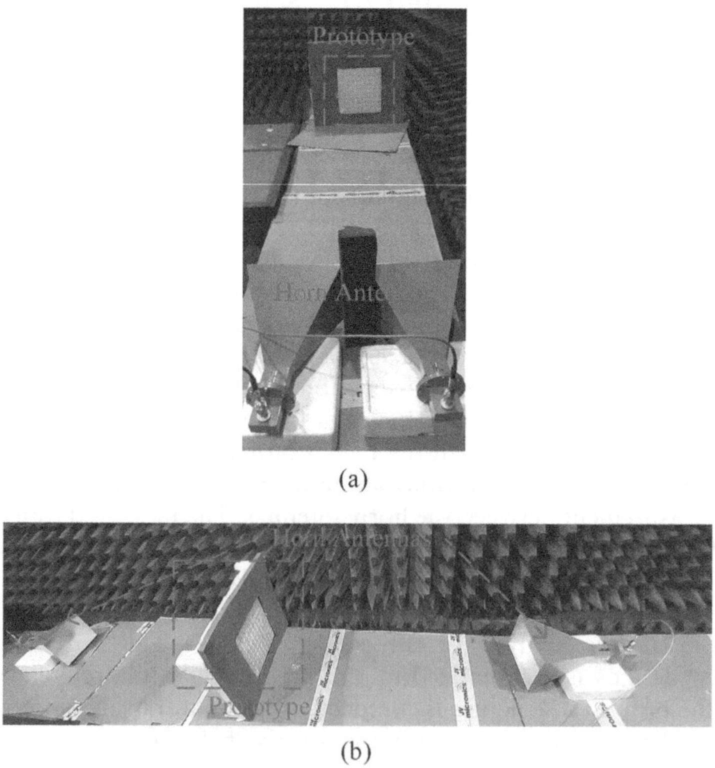

Figure 5.15 Measurement setups for (a) reflection and (b) transmission coefficients [1].

Table 5.3 Performance Comparison of the Proposed FSR [1]

Reference	A/T Location	Transmission Band/I.L.	Absorption Band	Size	Thickness	Polarization
[2]	T-A	1.5 GHz/1.5 dB	4.95 – 7.0 GHz (34.3%)	$0.001\lambda_L^2$	$0.0432\lambda_L$	Dual
[3]	A-T	10 GHz/0.2 dB	3 – 9 GHz (100%)	$0.072\lambda_L^2$	$0.12\lambda_L$	Dual
[4]	A-T-A	5.7 GHz /0.48 dB	2.60–5.25 GHz (67.51%) and 6.058.50 GHz (33.67%)	$0.054\lambda_L^2$	$0.134\lambda_L$	Dual
[5]	A-T-A	6 GHz / 0.5 dB	2.18–4.43 GHz (68.07%) and 7.559.98 GHz (27.72%)	$0.019\lambda_L^2$	$0.078\lambda_L$	Insensitive
[6]	A-T-A-T	7.2 GHz/2.30 dB and 13.05 GHz/1.69 dB	4.02–6.27 GHz (43.73 %) and 8.21–12.02 GHz (37.66 %)	$0.029\lambda_L^2$	$0.107\lambda_L$	Sensitive
This Work	T-A-T	2.1 GHz/1.5 dB and 8.8 GHz / 1.5 dB	3.6 – 7.4 GHz (69.09%)	$0.026\lambda_L^2$	$0.09\lambda_L$	Insensitive

Note: A is the absorption band; T is the transmission band; λ_L denotes the free space wavelength corresponding to the lowest frequency of −10 dB reflection; I.L. is the insertion loss at the transmission band.

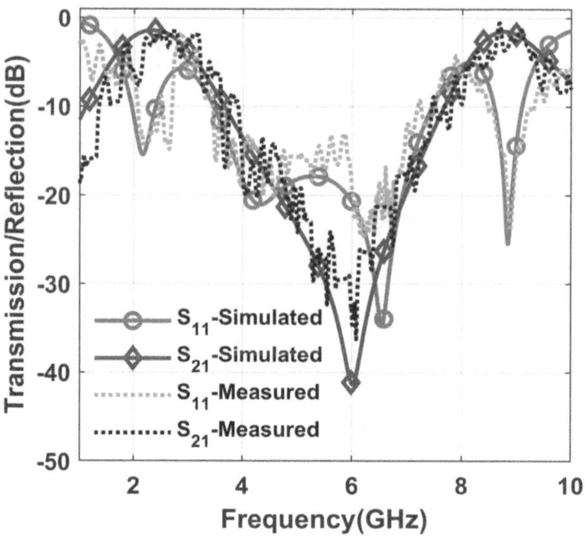

Figure 5.16 Simulated and measured reflection and transmission coefficients of the proposed FSR [1].

5.5 CONCLUSION

This chapter presents the development of a polarization-insensitive T-A-T-type rasorber featuring dual transmission bands. The design integrates a square-loop-based broadband absorber with a dual-band bandpass frequency selective surface (FSS), resulting in a wide absorption band (3.6–7.4 GHz) situated between two transmission bands centered at 2.1 GHz and 8.8 GHz. The operational mechanism is analyzed using an equivalent circuit model (ECM). A parametric study establishes the tunable range of transmission frequencies in relation to the absorption center frequency, based on the bandpass FSS parameters. A 10 × 10 array of the proposed FSR unit cells is fabricated, and experimental results are provided to validate its performance.

REFERENCES

1. M. M. Zargar et al., "Polarization-Insensitive Frequency Selective Rasorber with broad absorption band and two-sided transmission bands," Wiley International Journal of RF and Microwave Computer-Aided Engineering, vol. 32, no. 4, e23067, 2022
2. W. Yu et al., "Dual-Polarized Band-Absorptive Frequency Selective Rasorber Using Meander-Line and Lumped Resistors," IEEE Trans. Antenn. Propag., vol. 67, no. 2, pp. 1318–1322, Feb. 2019.

3. Q. Chen, S. Yang, J. Bai and Y. Fu, "Design of Absorptive/Transmissive Frequency-Selective Surface Based on Parallel Resonance," IEEE Trans. Antenn. Propag., vol. 65, no. 9, pp. 4897–4902, Sept. 2017.
4. M. Qu, S. Sun, L. Deng and S. Li, "Design of a Frequency-Selective Rasorber Based on Notch Structure," IEEE Access, vol. 7, pp. 3704–3711, 2019.
5. A. Sharma, M. Saikia, S. Malik, S. Ghosh and K. V. Srivastava, "A Polarization Insensitive Broadband Rasorber With in-Band Transmission Response," Microw. Opt. Technol. Lett., vol. 62, no. 12, pp. 3668–3676, 2020.
6. M. Guo, Q. Chen, Z. Sun, D. Sang and Y. Fu, "Design of Dual-Band Frequency-Selective Rasorber," IEEE Antenn. Wirel. Propag. Lett., vol. 18, no. 5, pp. 841–845, May 2019.

Chapter 6

A-T-A FSR for Low-RCS Radiating System

In the preceding chapters, A-T and T-A-T types of frequency selective rasorbers (FSRs) were explored. However, for effective suppression of out-of-band radar cross-section (RCS) in radiating systems, A-T-A configurations, featuring transmission within an absorption band on both sides, are more advantageous. This chapter focuses on designing a compact, polarization-insensitive A-T-A rasorber with a high-Q transmission band, tailored for electromagnetic shielding and RCS reduction in narrowband radiators. As a practical application, the proposed high-Q A-T-A FSR is integrated as a radome over a patch antenna operating at the same frequency as the rasorber's passband. This co-designed system is experimentally validated, demonstrating significant out-of-band RCS suppression while preserving the in-band radiating performance.

6.1 PROPOSED A-T-A FSR

A typical frequency selective rasorber (FSR) consists of two layers: a top resistive layer and a bottom layer comprised of a bandpass frequency selective surface (FSS). The resistive layer, integrated with lossy components, is responsible for absorbing incoming electromagnetic (EM) waves, while the bandpass FSS facilitates signal transmission. To achieve an absorption-transmission-absorption (A-T-A) response, the transmission band must be embedded within the broader absorption band. However, this overlap often results in increased insertion loss at the target transmission frequency. To mitigate this issue, an additional resonant element is incorporated into the lossy layer to enable transmission within the intended band by locally suppressing absorption.

The proposed FSR is fabricated on a Rogers Kappa substrate with a thickness of 0.8 mm, a dielectric constant (ϵ_r) of 4.3, a loss tangent ($\tan \delta$) of 0.005, and a copper cladding thickness of 35 μm. The design process involves the following key steps:

1. Initially, a broadband absorber is developed to ensure wideband electromagnetic absorption.

DOI: 10.1201/9781003644323-6

2. A bandpass FSS is then analyzed and designed to operate at the target transmission frequency.
3. This bandpass FSS replaces the ground plane of the broadband absorber to form the FSR structure.
4. An MF-shaped high-Q resonator is integrated with the lossy layer to enable narrowband transmission within the absorption band.
5. The lossy layer is further optimized to enhance absorption at higher frequencies and improve overall performance.

6.1.1 Design and Analysis

The FSS-based broadband absorber employs a dual-layer configuration comprising a resistive surface and a ground plane, as illustrated in Figure 6.1. The resistive surface features a metallic square loop connected to rectangular strips on all four sides, as shown in Figure 6.1(b). Surface-mount chip resistors with a resistance value of 200 Ω are positioned within the gaps of these strips. Electromagnetic performance is evaluated using CST Microwave Studio. The separation between the resistive layer and the ground plane is set to a quarter-wavelength ($\lambda/4$), with λ corresponding to the center frequency of the intended absorption band. As indicated by the reflection coefficient in Figure 6.1(c), values remain below −10 dB across the 4.2–12.7 GHz frequency span, translating to over 90% absorption efficiency in this range.

To explore transmission characteristics, a bandpass frequency selective surface (FSS) is developed by etching a double cross-slot (DCS) pattern into a metal-coated substrate, as shown in Figure 6.2(a). The simulated S-parameter response in Figure 6.2(b) confirms a passband centered around 8.3 GHz. The metallic ground plane of the absorber is then replaced with this DCS-based bandpass FSS, forming the structure presented in Figure 6.3(a). The updated configuration's response, shown in Figure 6.3(b), reveals considerable insertion loss (exceeding 5 dB) at the passband frequency, indicating inefficient transmission. This poor performance is further exacerbated by reflection levels above −10 dB. The excessive insertion loss at the transmission frequency is primarily attributed to its placement within the absorber's operating band, where energy dissipation is elevated due to the lossy components integrated into the resistive surface.

To address the excessive losses at the transmission frequency, an MF-shaped resonator offering sharper transmission characteristics is embedded within the central region of the square loop resistive structure, as illustrated in Figure 6.4. A cross-shaped slot is introduced into the MF-resonator to enhance its performance. The geometry and dimensions of the resonator are optimized to generate a transmission pole precisely aligned with the desired frequency located within the broader absorption band. At resonance, the surface currents become concentrated along the

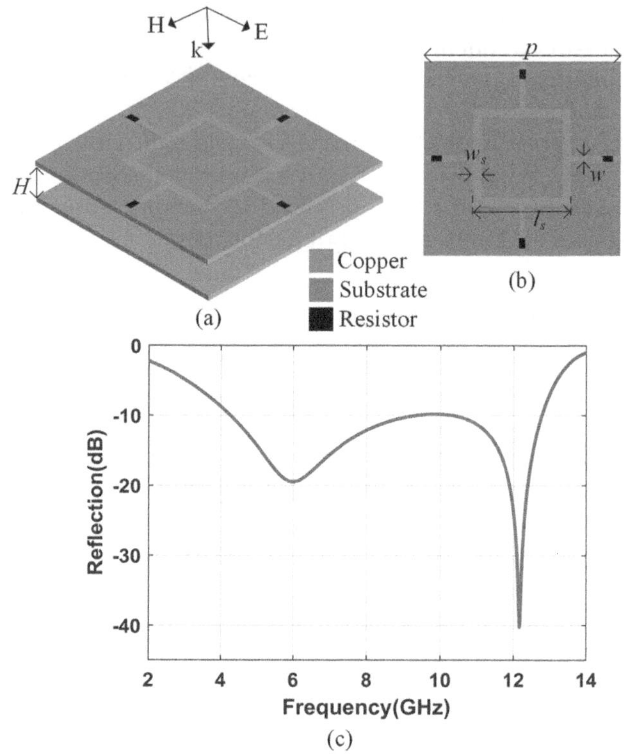

Figure 6.1 (a) Perspective, (b) front view of the unit cell geometry, and (c) simulated reflection coefficient of the broadband absorber: p = 15, w = 0.5, w_s = 0.8, l_s = 7.1, H = 8 (all dimensions are in millimeters) [1].

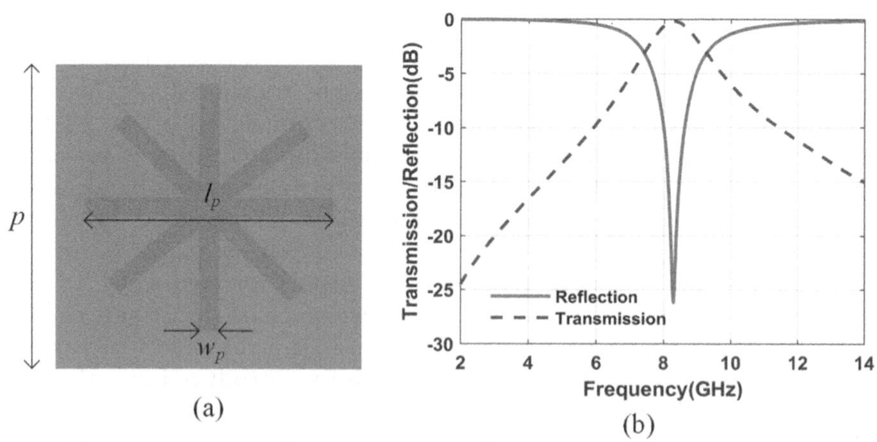

Figure 6.2 (a) Front view of unit cell geometry and (b) simulated reflection/transmission coefficients of the bandpass FSS: p = 15, l_p = 12, w_p = 0.6 (all dimensions are in millimeters) [1].

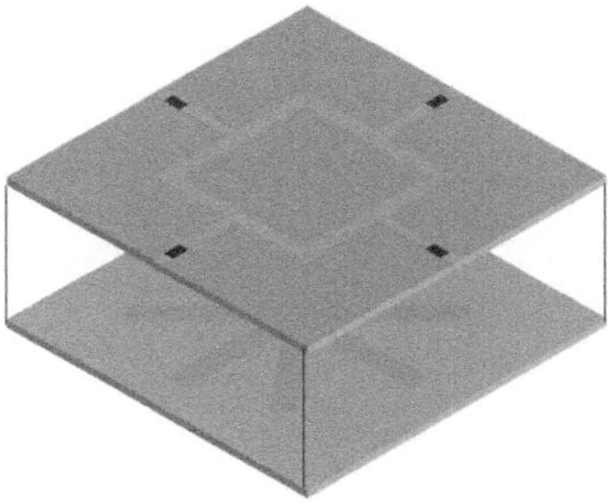

(a)

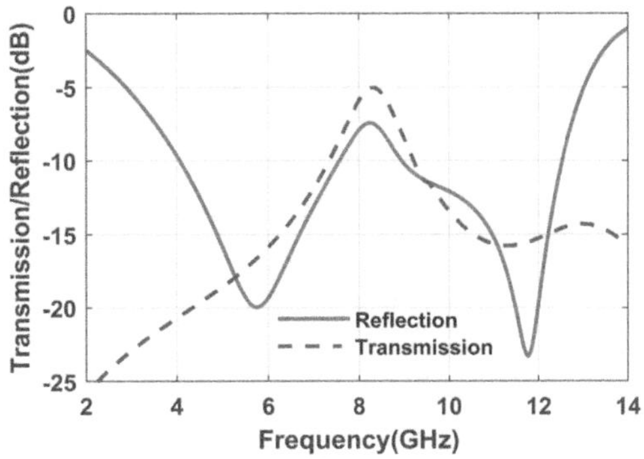

(b)

Figure 6.3 (a) Perspective view of the unit cell geometry and (b) simulated reflection/transmission coefficients of the square loop resistive layer in combination with the bandpass FSS layer [1].

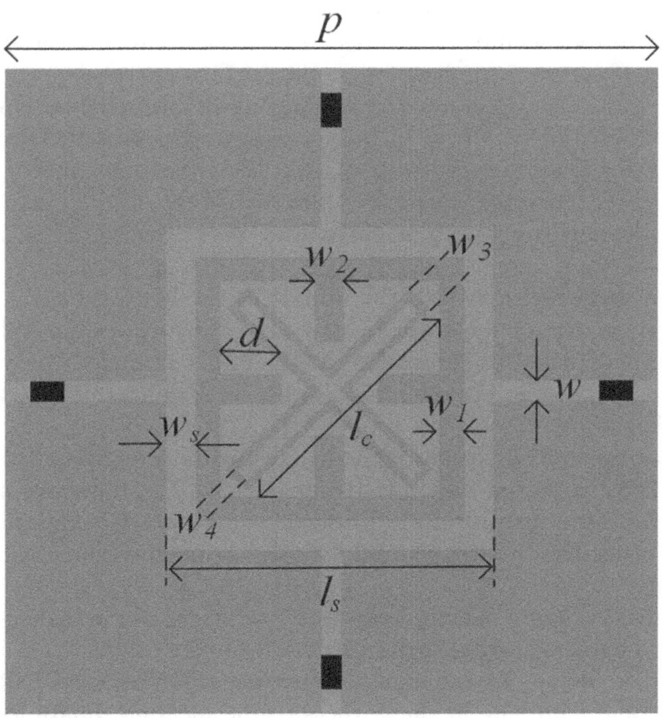

Figure 6.4 Front view of the unit cell with MF-shaped resonator: $p = 15$, $w = 0.5$, $w_s = 0.8$, $l_s = 7.1$, $w_1 = 0.3$, $w_2 = 0.3$, $l_c = 5.8$, $w_3 = 0.6$, $w_4 = 0.2$, $d = 1.6$ (all dimensions are in mm) [1].

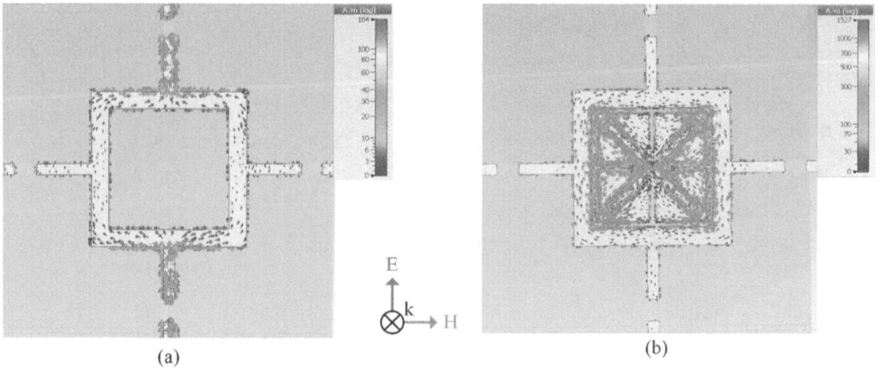

Figure 6.5 Surface current distribution of the resistive layer at 8.3 GHz: (a) without and (b) with MF resonator [1].

etched slots of the MF-resonator, effectively diverting the flow from the resistive components and thereby minimizing power loss. This behavior is clarified by examining the surface current profiles at the transmission frequency in two scenarios, as presented in Figure 6.5. Without the MF-resonator, Figure 6.5(a) shows that the surface currents predominantly flow along the rectangular strip, forcing significant current through the chip resistors and resulting in higher losses. In contrast, Figure 6.5(b) demonstrates that the addition of the MF-shaped resonator redirects the resonance currents along the slot features, significantly reducing current through the resistive paths and thus lowering the associated ohmic losses.

To gain deeper insight into the impact of the MF resonator, the simulated performance of the front resistive layer incorporating only the MF resonator is depicted in Figure 6.6. At the target frequency of 8.3 GHz, a noticeable decrease in insertion loss is observed, indicating effective transmission behavior.

In the complete FSR structure, the MF resonator integrated with the lossy resistive layer is paired with the bandpass FSS. The full-wave simulation results of this configuration are illustrated in Figure 6.7, revealing a pronounced transmission peak around 8.4 GHz situated within a wide absorption range.

As shown in Figure 6.7, the reflection coefficient in the higher-frequency absorption band slightly exceeds −10 dB, leading to a reduction in absorptivity in that region. To enhance performance, the proposed FSR incorporates diagonal trimming at the corners of the metallic square loop on the resistive layer, as illustrated in Figure 6.8. The impact of this design adjustment is evident in Figure 6.9, where the reflection coefficient in the upper absorption band drops below −10 dB. This indicates an improvement in absorption efficiency at higher frequencies due to the corner modification of the front square loop.

The designed A-T-A-type FSR features a distinct transmission window centered at 8.4 GHz, flanked by absorption bands on either side. The lower absorption band, offering more than 80% absorption, extends from 3.4 GHz to 7.5 GHz, corresponding to a fractional bandwidth of 75.22%. The

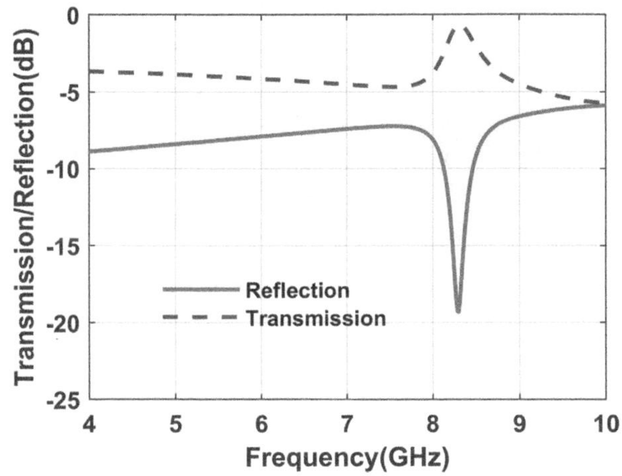

Figure 6.6 Simulated coefficients for the front resistive layer only with MF resonator [1].

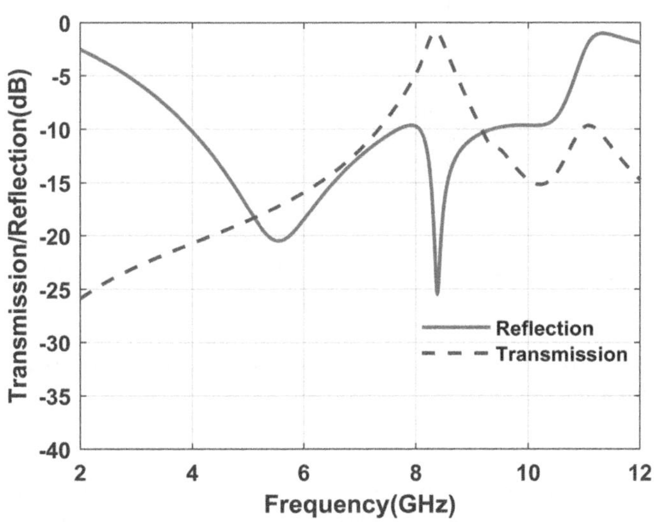

Figure 6.7 Simulated transmission and reflection coefficients of the proposed FSR [1].

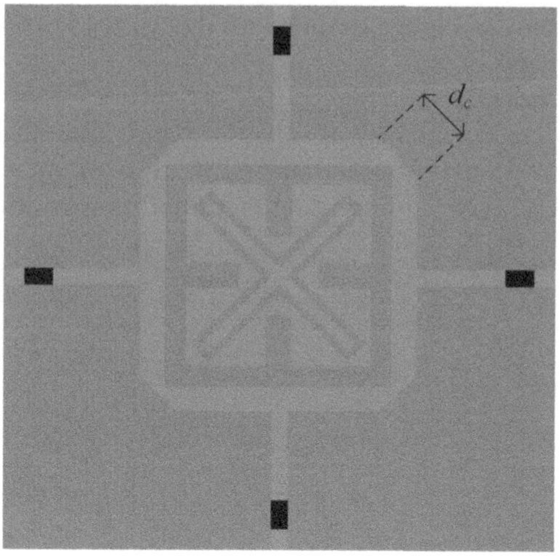

Figure 6.8 Front view of the proposed FSR unit cell with corner trimmed square loop [1].

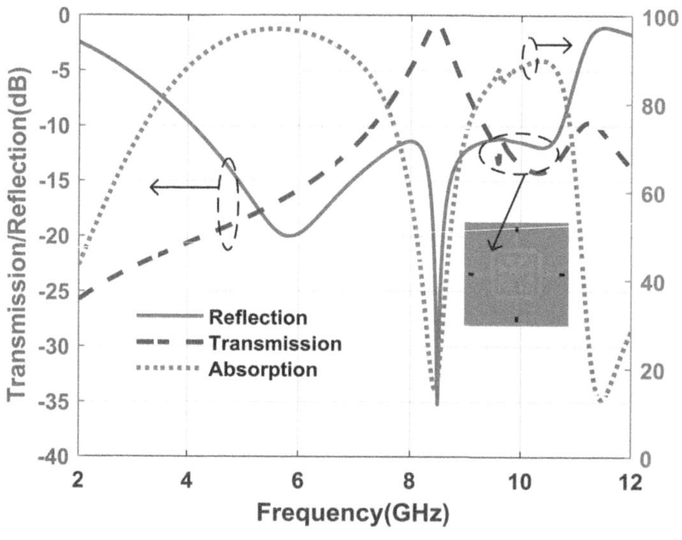

Figure 6.9 Simulated response of the proposed FSR with corner trimmed square loop [1].

upper absorption region spans from 9.2 GHz to 10.8 GHz, covering 16% bandwidth. At the center transmission frequency of 8.45 GHz, the structure exhibits a low insertion loss of 0.7 dB. The 3 dB bandwidth for the transmission band ranges from 8.17 GHz to 8.69 GHz. The quality factor (Q) of the transmission band, calculated as the ratio of center frequency to its 3 dB bandwidth, is 16.25, indicating high frequency selectivity of the proposed FSR.

The performance of the proposed FSR is also evaluated under varying polarization angles of the incident electromagnetic wave. Figure 6.10(a) illustrates the simulated results for different polarization orientations. The

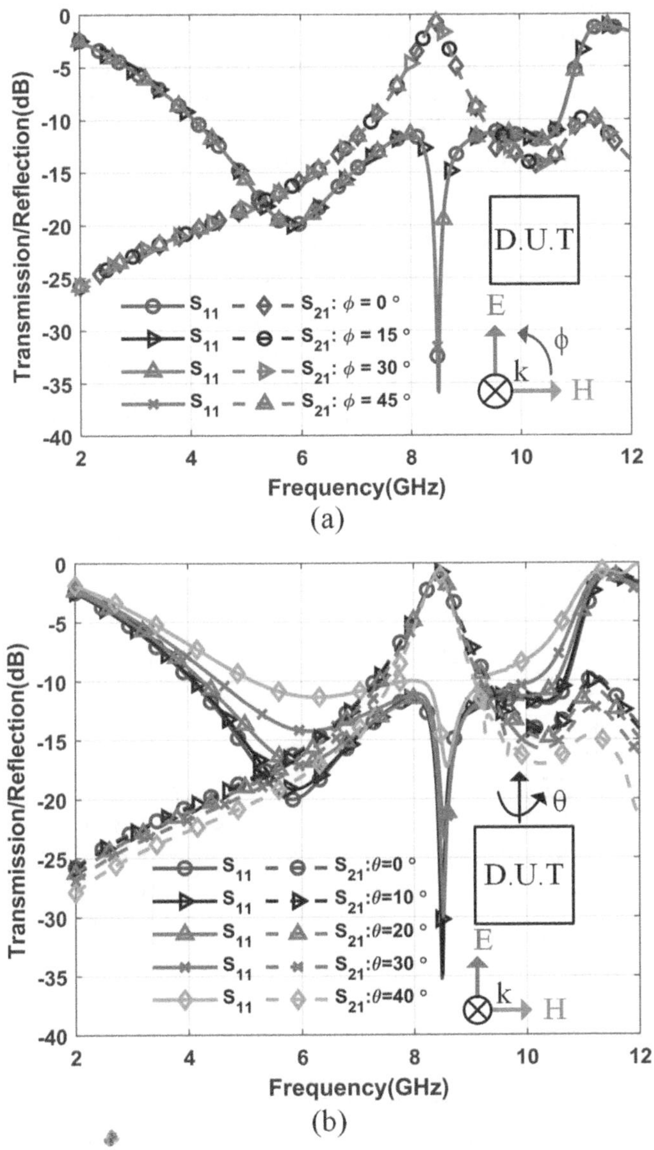

Figure 6.10 Simulated transmission and reflection coefficients of the proposed FSR under various (a) polarizations and (b) incidence angles [1].

transmission and reflection characteristics show negligible variation with changes in polarization, indicating stable behavior. This consistent response can be attributed to the four-fold symmetric design of the FSR, confirming its polarization-insensitive nature.

The proposed FSR is further evaluated for performance under oblique incidence conditions. As shown in Figure 6.10(b), its response is analyzed for incident angles up to 40°. With increasing angle, a gradual reduction in absorptivity is observed in both the lower and upper absorption bands, whereas the transmission band remains relatively unaffected. Despite the drop in absorption, the performance on either side of the transmission band remains within acceptable limits up to a 40° incidence. This indicates that the proposed FSR maintains stable operation for oblique angles up to 40°.

6.1.2 Equivalent Circuit Model

The operation of the proposed FSR is further examined through its equivalent circuit model (ECM). When an electromagnetic wave impinges on the FSR surface, it induces surface currents on the metallic patterns. These currents give rise to distributed inductive and capacitive effects, which are represented using lumped inductors and capacitors in the ECM. Figure 6.11

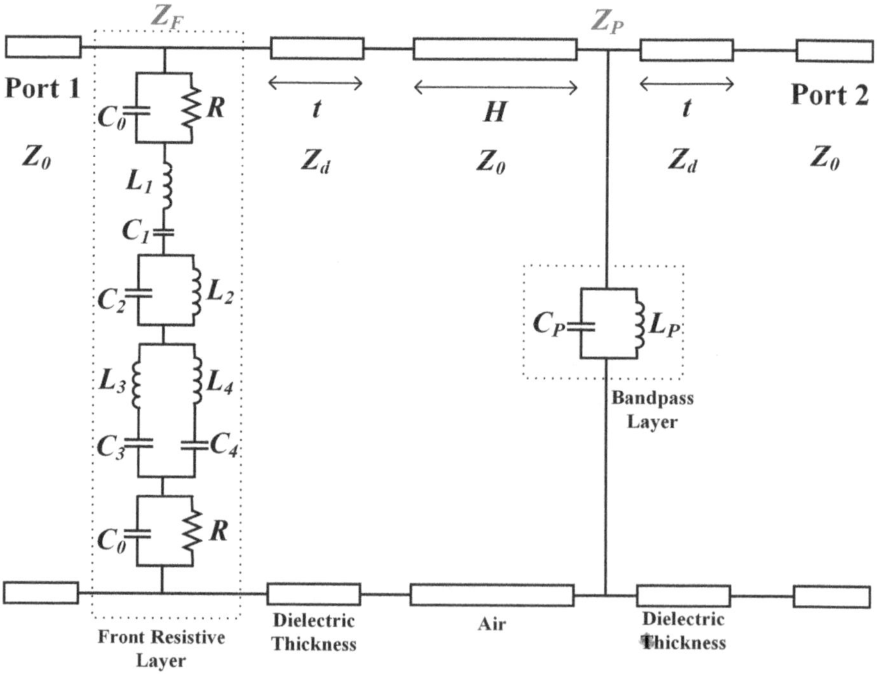

Figure 6.11 ECM corresponding to the proposed FSR [1].

presents the developed ECM, where the equivalent representations of the resistive layer, dielectric substrate, air gap, and bandpass FSS are connected in a cascaded configuration to mimic the overall FSR behavior.

The mounted lumped resistor and the associated gap capacitance are represented by a parallel R–C_0 network. The inductance of the rectangular strip adjacent to the diagonally trimmed square loop is modeled using inductor L_1. The metallic square loop resonator is characterized by a parallel L_2–C_2 circuit. The MF-shaped resonator, incorporating a cross-loop slot, is represented by two series L_3–C_3 and L_4–C_4 branches in parallel. The coupling capacitance between the square loop and the MF-shaped resonator is denoted by C_1. The double cross-slot structure forming the bandpass layer is modeled with a parallel L_P–C_P configuration. The air gap separating the layers is simulated using a transmission line with a characteristic impedance of Z_0 (free space) and length H. The top and bottom dielectric substrates are modeled as transmission lines with characteristic impedance $Z_0/\sqrt{\epsilon_r}$ and length equal to the substrate thickness (t). The entire equivalent circuit behaves as a two-port network matched to free space at both ports.

The ABCD parameters of the equivalent circuit model (ECM) are determined by successively multiplying the ABCD matrices of each individual layer, as expressed in the following equation:

$$\begin{bmatrix} A & B \\ C & D \end{bmatrix} = \begin{bmatrix} 1 & 0 \\ 1/Z_\mathrm{F} & 1 \end{bmatrix} \times \begin{bmatrix} \cos\beta t & jZ_d\sin\beta t \\ j\sin\beta t / Z_\mathrm{d} & \cos\beta t \end{bmatrix}$$
$$\times \begin{bmatrix} \cos\beta_0 H & jZ_0\sin\beta_0 H \\ j\sin\beta_0 H / Z_0 & \cos\beta_0 H \end{bmatrix} \times \begin{bmatrix} \cos\beta t & jZ_d\sin\beta t \\ j\sin\beta t / Z_\mathrm{d} & \cos\beta t \end{bmatrix} \quad (6.1)$$
$$\times \begin{bmatrix} 1 & 0 \\ 1/Z_\mathrm{P} & 1 \end{bmatrix}$$

where propagation constant β is calculated as $\beta = \beta_0\sqrt{\mu_r\epsilon_r}$ (μ_r and ϵ_r denote the relative permeability and permittivity of the dielectric substrate, respectively). The impedances of the top resistive layer and the bottom bandpass FSS, represented by Z_F and Z_P, respectively, are defined in Equations (6.2) and (6.3).

$$Z_F = \frac{2R}{1-\omega^2R^2(C_0)^2} + j\left\{\omega L_1 - \frac{1}{\omega C_1} - \frac{2R^2\omega C_0}{1-\omega^2R^2(C_0)^2}\right.$$
$$\left. + \frac{\omega L_2}{1-\omega^2C_2L_2} - \frac{(1-\omega^2L_3C_3)(1-\omega^2L_4C_4)}{\omega C_3 + \omega C_4 - \omega^3L_3C_3C_4 - \omega^3L_4C_3C_4}\right\} \quad (6.2)$$

$$Z_P = \frac{j\omega L_P}{1-\omega^2L_PC_P} \quad (6.3)$$

The S-parameters of the equivalent circuit model (ECM) are derived from the ABCD parameters using the conversion formulas:

$$S_{11} = \frac{A + B/Z_0 - CZ_0 - D}{A + B/Z_0 + CZ_0 + D} \tag{6.4}$$

$$S_{21} = \frac{2}{A + B/Z_0 + CZ_0 + D} \tag{6.5}$$

The absorptivity $A(\omega)$ is then determined using the S-parameters derived in Equations (6.4) and (6.5), as expressed in Equation (6.6).

$$A(\omega) = 1 - |S_{11}|^2 - |S_{21}|^2 \tag{6.6}$$

The equivalent circuit model (ECM) developed for the proposed FSR is evaluated using the Keysight ADS simulation tool. The optimum value of circuit components are given as $R = 200\ \Omega$, $C_0 = 0.036085\ \text{pF}, C_1 = 0.187\ \text{pF}, L_1 = 2.45\ \text{nH}, L_2 = 0.53\ \text{nH}, C_2 = 0.338\ \text{p}$ $L_3 = 0.33\ \text{nH}, C_3 = 0.401\ \text{pF}, L_4 = 1.085\ \text{nH}, C_4 = 0.65\ \text{pF}, L_P = 0.7\ \text{nH}$, $C_P = 0.5\ \text{pF}, Z_d = 181\ \Omega, t = 0.8\ \text{mm}, Z_0 = 377\ \Omega)$. A comparison between the ECM's S-parameter results and the full-wave simulation of the FSR is shown in Figure 6.12. The reflection and transmission profiles derived from the ECM demonstrate a strong correlation with the A-T-A response observed in the full-wave analysis, particularly showcasing a central transmission

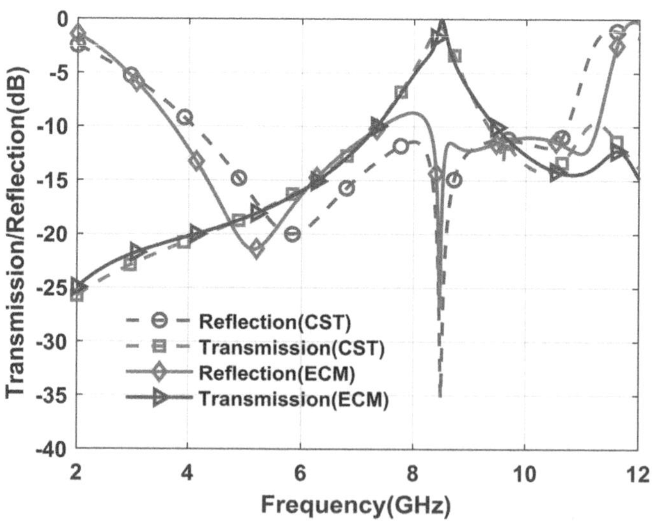

Figure 6.12 Simulated transmission and reflection coefficients of the ECM and full-wave simulation of the proposed FSR [1].

peak at 8.4 GHz flanked by absorption bands. This confirms that the ECM effectively captures the functional behavior of the proposed FSR.

6.2 EXPERIMENTAL MEASUREMENT OF THE PROPOSED FSR AND DISCUSSIONS

To experimentally verify the performance of the proposed FSR, measurements were conducted on a fabricated prototype. This prototype comprises a 17 × 17 array of unit cells constructed on a 0.8 mm thick Rogers Kappa substrate, characterized by a dielectric constant (ϵ_r) of 4.3 and a loss tangent (tan δ) of 0.005. Photographs of the complete prototype, measuring 255 mm × 255 mm, are shown in Figure 6.13. The top resistive layer, which includes mounted surface-mount chip resistors (CRCW0603200RFKEA from VISHAY) with a resistance of 200 Ω, is displayed in Figure 6.13(a). Figure 6.13(b) illustrates the bottom bandpass FSS layer featuring a double cross-slot (DCS) pattern. The full assembly of the two-layer FSR structure, incorporating plastic spacers to separate the resistive and FSS layers, is presented in Figure 6.13(c).

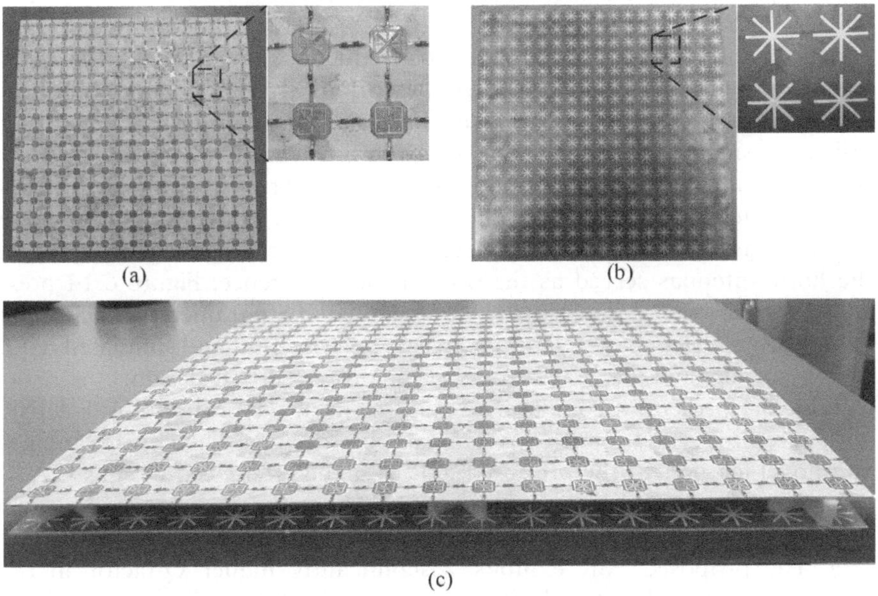

Figure 6.13 Photographs of the fabricated prototype of the proposed FSR: (a) top layer, (b) bottom layer, and (c) side view [1].

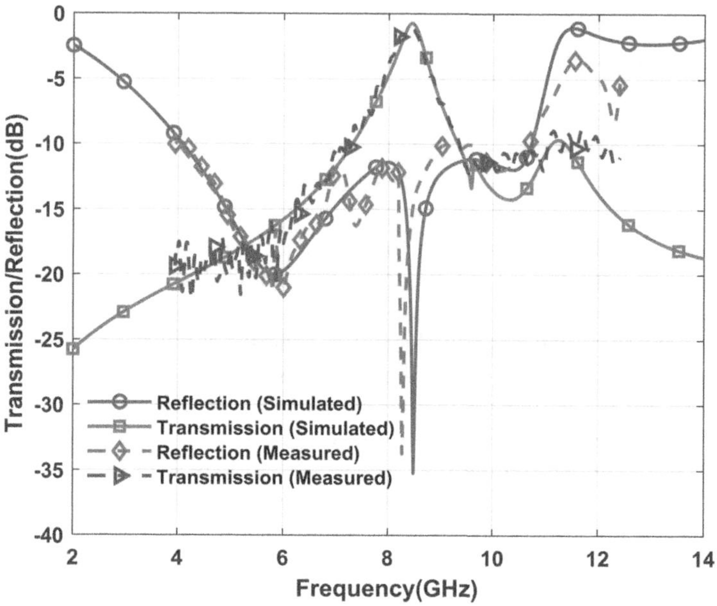

Figure 6.14 Measured transmission and reflection coefficients of the fabricated 17 × 17 array of FSR prototype compared with the simulated results [1].

Free-space measurements were performed separately for the C (3.95–5.85 GHz), J (5.85–8.20 GHz), and X (8.20–12.4 GHz) bands using standard gain horn antennas appropriate for each frequency range. The S-parameters were recorded using a Keysight PNA Network Analyzer N5224B. For reflection measurements, the time-domain feature of the analyzer was employed, utilizing a single horn antenna and applying time gating to isolate the reflection from the fabricated sample. Both reflection and transmission coefficients were measured with respect to their respective references: a metallic plate was used as the reflection reference, while the direct signal path between the horn antennas served as the transmission reference. Figure 6.14 presents a comparison of the measured and simulated S-parameters. The results demonstrate a strong correlation between experimental and simulated data, thereby confirming the effectiveness of the proposed FSR design.

Table 6.1 presents a comparative assessment of the proposed A-T-A FSR with several recently reported A-T-A rasorber designs. The key advantages and novel features of the proposed FSR over the referenced designs [2–13] are summarized as follows:

- The proposed FSR exhibits a significantly higher Q-factor in the transmission band, indicating superior selectivity compared to all the referenced designs. This high selectivity makes it highly suitable for shielding applications and electromagnetic interference (EMI) mitigation in narrowband radiating systems.

Table 6.1 Proposed A-T-A FSR in Comparison with the Recently Reported A-T-A Rasorbers [1].

Ref.	*Passband (I.L.)*	*Q-Factor*	*Lower Absorption B.W. (GHz)*	*Upper Absorption B.W.(GHz)*	*Thickness*	*Layers*	*Unit Cell Size*	*RCS Reduction Verified*
[2]	5.7 GHz (0.48 dB)	12.95	2.6–5.25 (67.5%)	6.05–8.5 (33.7%)	15.524 mm (0.15λ_L)	2	27 × 27 mm² (0.075λ_L^2)	No
[3]	5 GHz (0.35 dB)	7.35	2.76–4.40 GHz (45.81%)	5.64–8.42 (39.5%)	11 mm (0.101λ_L)	2	14 mm Hexagon (0.037λ_L^2)	No
[4]	3.63 GHz (0.14 dB)	6.36	1.9–2.74 (57.7%)	4.04–6.82 (51.1%)	13 mm (0.082λ_L)	2	25 x 25 mm² (0.024λ_L^2)	No
[5]	6.5 GHz (0.67 dB)	DNG	1.5–4.6 (101.63%)	9.2–13.7 (39.3%)	32 mm (0.16λ_L)	4	20 × 20 mm² (0.01λ_L^2)	No
[6]	6.01 GHz (0.38 dB)	DNG	2.95–5.65 (62.7%)	7.4–9.24 (22.1%)	6.5 mm(0.063λ_L)	2	20 × 20 mm² (0.038λ_L^2)	No
[7]	6.6 GHz (0.44 dB)	6.05	2.89–5.59 (63.8%)	8.79–8.85 (6.03%)	10.17 mm(0.097λ_L)	4	20 × 20 mm² (0.037λ_L^2)	No
[8]	8.7 GHz (1 dB)	3.28	– (63.51%)	– (31.84%)	9.802 mm(0.113λ_L)	4	10 × 10 mm² (0.014λ_L^2)	No
[9]	10.4 GHz (0.52 dB)	DNG	6.26–9.58 (41.91%)	11.46–14.06 (20.37%)	6.016 mm (0.125λ_L)	2	18 × 18 mm² (0.141λ_L^2)	No
[10]	7.9–9.0 GHz (1 dB)	7.68	3.3–7.1 (73.07%)	9.4–12.0 (24.29%)	10.45 mm(0.01λ_L)	4	18 × 18 mm² (0.031λ_L^2)	No
[11]	6.3 GHz (0.78 dB)	DNG	2.3–5.3 (78.94%)	7.8–14.6 (60.71%)	13.85 mm(0.111λ_L)	3	20 × 20 mm² (0.023λ_L^2)	No
[12]	10.28 GHz (0.47 dB)	DNG	DNG	DNG	11.93 mm(0.122λ_L)	3	9 × 18 mm²(0.016λ_L^2)	No
[13]	8–12.5 GHz (2 dB)	DNG	1.5–3.6 (82.35%)	15.2–21.8 (35.67%)	23 mm(0.126λ_L)	3	60 × 60 mm² (0.09λ_L^2)	No
This Work	**8.4 GHz (0.7 dB)**	**16.25**	**3.4–7.5 (75.22%)**	**9.2–10.8 (16%)**	**9.6 mm** (0.108λ_L)	**2**	**15 × 15 mm²** (0.028λ_L^2)	**Yes**

Note: λ_l =free space wavelength for the lowest absorption frequency; DNG = data not given.

- In terms of absorption, the proposed structure achieves a broader lower-band absorption of 75.22%, outperforming the designs in [2–4], [6–10], and offers a 16% absorption bandwidth in the upper band, which surpasses the performance of [7].
- Unlike several previously reported designs that employ 3-layer [11–13] or 4-layer [5, 7, 8, 10] structures, the proposed FSR uses only two layers, making the design more compact and fabrication-friendly.
- The unit cell dimensions in the proposed FSR are smaller than those in works such as [2, 3, 6, 7, 9, 10, 13], contributing to better spatial efficiency.
- The overall thickness of the proposed structure is reduced compared to designs in [2, 5, 8, 9, 11–13], enhancing integration with antenna systems.
- Furthermore, the proposed FSR has been experimentally validated as a radome for a narrowband patch antenna, demonstrating its effectiveness in out-of-band RCS suppression – an application not shown in the compared works [2–13].

6.3 LOW-RCS ANTENNA CO-DESIGNED WITH PROPOSED FSR

This section presents the integration of the proposed A-T-A rasorber with a narrowband patch antenna to achieve a radiating system with suppressed out-of-band radar cross-section (RCS). The patch antenna is designed using a 0.762 mm thick Rogers Diclad substrate ($\varepsilon_r = 2.2$, $\tan \delta = 0.0009$), with its resonant frequency aligned with the transmission band of the proposed FSR. The patch's ground plane is sized at 75 mm × 75 mm, matching the footprint of a 5 × 5 array of the proposed FSR unit cells. As illustrated in Figure 6.15, the FSR array is positioned above the antenna at an approximate $\lambda/2$ spacing from the radiating surface, relative to the antenna's operational frequency. A comparative analysis of RCS and key antenna performance metrics is carried out between the FSR-loaded configuration and a conventional patch antenna. The fabricated assembly of the FSR-integrated antenna, comprising a 5 × 5 unit cell array placed above the patch, is depicted in Figure 6.16.

Figure 6.17 presents the simulated and experimental results of S_{11} and gain for both the standalone and FSR-integrated antenna. Both configurations exhibit an operating frequency near 8.4 GHz. At this frequency, the FSR-loaded antenna achieves a gain of 6.072 dBi, which is marginally lower than the 6.854 dBi gain observed for the conventional antenna. The radiation patterns of both antennas at 8.4 GHz in the *xz*- and *yz*-planes, as shown in Figure 6.18, indicate that the FSR-integrated antenna maintains a stable radiation profile, closely matching that of the reference antenna without significant pattern distortion.

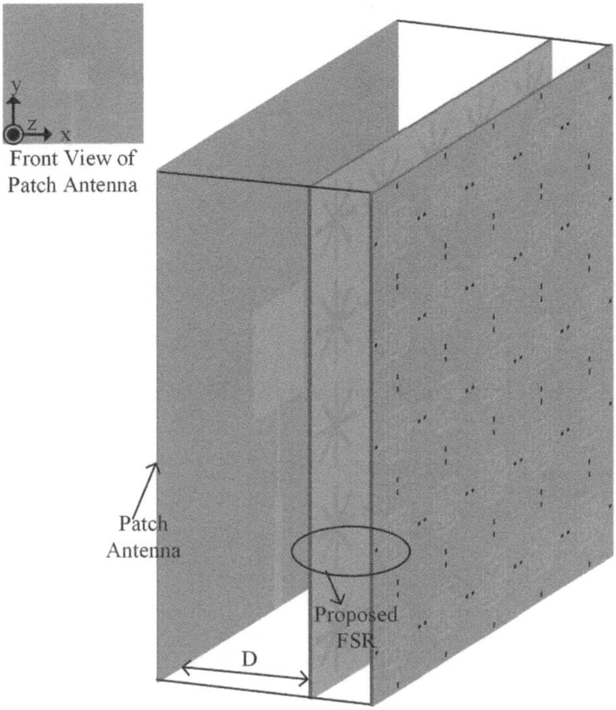

Figure 6.15 Schematic of the FSR-integrated patch antenna. D = 18 mm [1].

Figure 6.16 Fabricated prototype photograph of the FSR-integrated patch antenna [1].

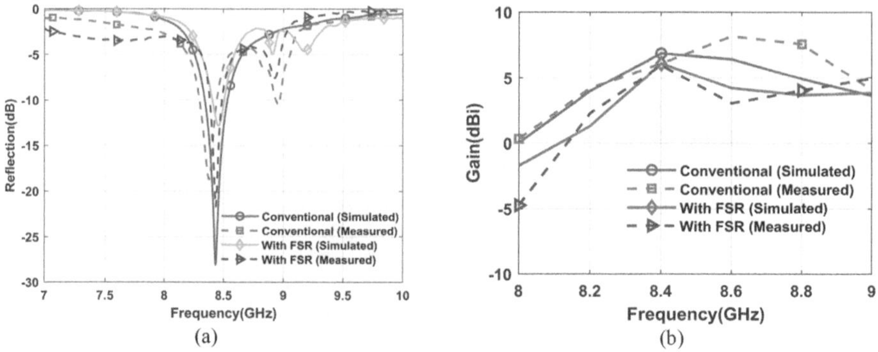

Figure 6.17 (a) Reflection coefficient and (b) gain of the patch antenna [1].

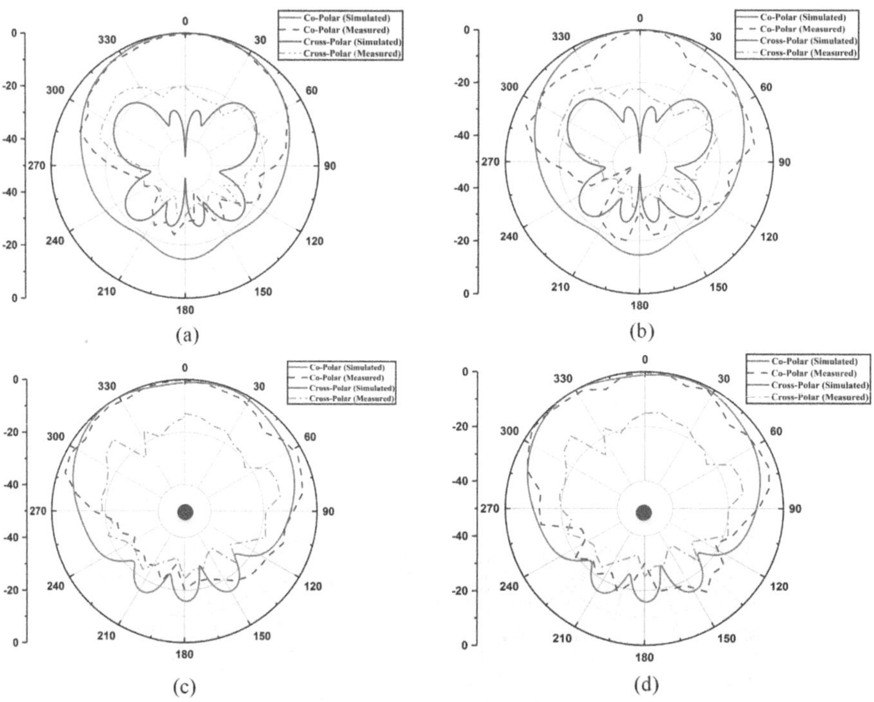

Figure 6.18 Radiation patterns of the patch antenna: (a) *xz*-plane (without FSR), (b) *xz*-plane (with FSR), (c) *yz*-plane (without FSR), and (d) *yz*-plane (with FSR) [1].

6.3.1 Monostatic RCS Measurement

RCS measurements were performed on the fabricated prototypes of the conventional antenna and the antenna integrated with the FSR. For monostatic RCS evaluation, the transmitting and receiving antennas share the same location. The monostatic RCS was measured using the time-domain gating feature of the Keysight PNA Network Analyzer N5224B. In this setup, a single horn antenna is employed to function as both transmitter and receiver. The RCS is determined using the standard expression provided in Equation (6.7) [14].

$$RCS = \frac{P_r}{P_t} \cdot \frac{(4\pi)^3 R^4}{G_t G_r \lambda^2} = K \cdot \frac{P_r}{P_t} \tag{6.7}$$

where P_r and P_t represent the power received and transmitted, respectively. G_r and G_t denote the gains of the receiving and transmitting antennas. R is the distance between the antennas and the target, and λ stands for the wavelength of the operating signal.

The monostatic RCS measurements for both the FSR-integrated antenna and the conventional antenna were performed using a reference metallic square PEC, matching the ground plane dimensions of the antenna. Figure 6.19 presents a comparative analysis of the monostatic RCS results. It is evident that the FSR-integrated antenna exhibits a notable reduction in RCS at out-of-band frequencies compared to the conventional design. However, near the antenna's resonant frequency, both configurations display similar RCS levels. In the lower frequency range, a peak RCS reduction of 23 dB is observed at 5.1 GHz, while in the upper range, the highest reduction reaches 10 dB at 10.9 GHz. On average, the FSR-integrated antenna achieves RCS reductions of 11.92 dB and 5.04 dB within the 4–7.5 GHz and 9.2–10.8 GHz bands, respectively.

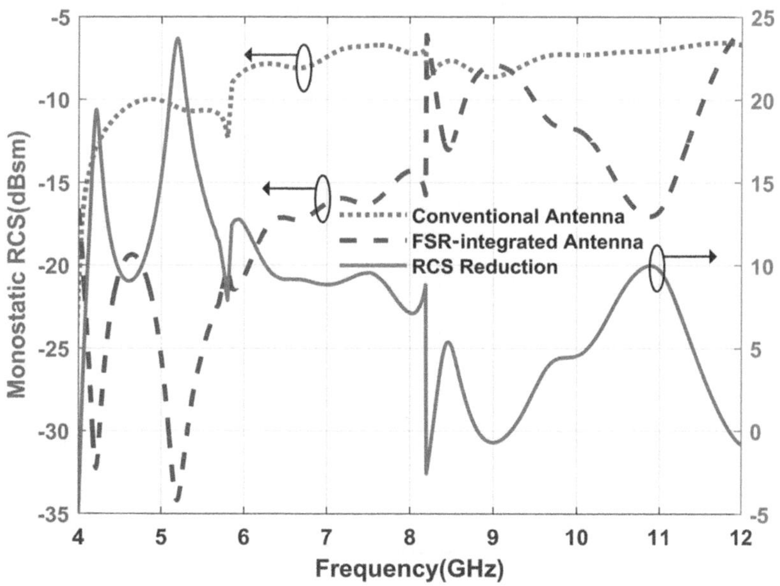

Figure 6.19 Monostatic RCS of the FSR-integrated antenna compared to the conventional patch antenna [1].

6.3.2 Bistatic Measurement

For bistatic RCS measurement, the transmitting and receiving horn antennas are positioned separately. As illustrated in Figure 6.20, the transmitting antenna remains fixed, directing the incident wave normally onto the object under test (OUT). Meanwhile, the receiving antenna is located at a far-field distance R from the OUT. It is rotated along a circular arc ranging from 0° to 90° in 10° increments, either clockwise or counterclockwise. At each angular position, the transmission coefficient is recorded using a PNA. The scattered electric field $E_{sca}(f,\phi)$ at each angle is expressed by the following equation [15]:

$$E_{sca}(f,\phi) = S_{21,O}(f,\phi) - S_{21,FS}(f,\phi) \tag{6.8}$$

where $S_{21,O}$ represents the transmission coefficient measured between the transmitting and receiving horn antennas with the object present, while $S_{21,FS}$ denotes the corresponding transmission coefficient in free space (i.e., without the object).

The normalized total radar cross-section $RCS_{t,\text{norm}}$ for the FSR-integrated antenna is evaluated by integrating the scattered field intensity, as described in Equation (6.9) [14].

$$RCS_{t,\text{norm}} = \frac{\int \left|E_{\text{sca},FSRA}(f,\phi)\right|^2 d\phi}{\int \left|E_{\text{sca},CA}(f,\phi)\right|^2 d\phi} \tag{6.9}$$

where $E_{\text{sca},FSRA}$ and $E_{\text{sca},CA}$ are the scattered fields corresponding to FSR-integrated and conventional antenna, respectively.

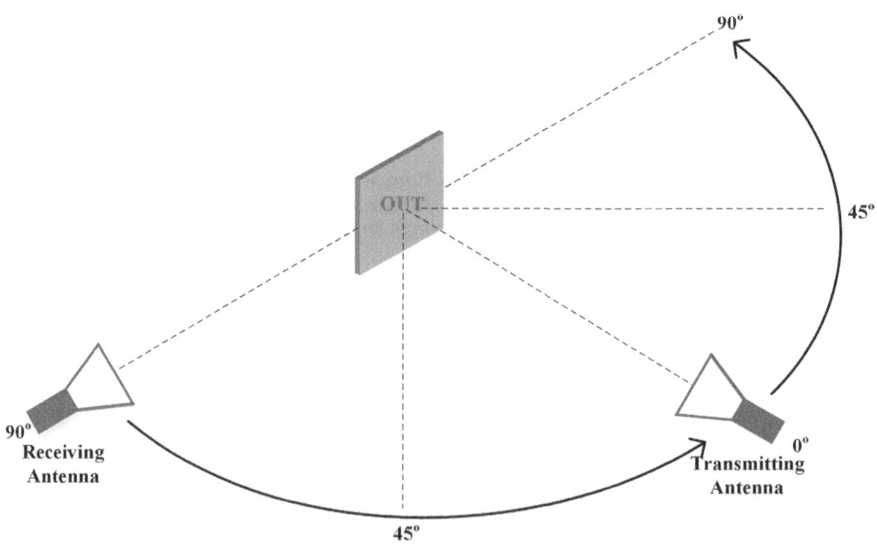

Figure 6.20 Schematic for the bistatic measurement setup [1].

Figure 6.21 illustrates the normalized total radar cross-section $RCS_{t,\text{norm}}$ of the FSR-integrated antenna in comparison with the conventional antenna. A noticeable reduction in $RCS_{t,\text{norm}}$ is observed for the FSR-based design. The lowest values achieved are 0.02 at 4.1 GHz and 0.15 at 10.4 GHz, corresponding to reductions of 98% and 85% in total bistatic RCS, respectively, compared to the conventional antenna. The average $RCS_{t,\text{norm}}$ is calculated to be 0.23 in the lower band (4–7.5 GHz) and 0.40 in the upper band (9.2–10.8 GHz), leading to average bistatic RCS reductions of approximately 77% and 60% in the respective absorption bands.
Table 6.2 presents a performance comparison between the proposed FSR-integrated antenna and other recently reported low-RCS radome-based antenna designs. Key highlights of the proposed design include:

1. The FSR-integrated antenna exhibits a more compact footprint compared to the designs discussed in [18, 20], and it also features a lower electrical thickness than those presented in [18, 19].
2. It offers reduced structural complexity, using fewer substrate layers than the antennas in [17, 18], and requires fewer lumped elements per unit cell than the designs in [16, 18, 19].
3. Unlike the antennas in [16–20], the proposed FSR-integrated antenna has been experimentally validated to achieve substantial average total bistatic RCS reductions of 77% in the 4–7.5 GHz range and 60% in the 9.2–10.8 GHz range, when compared to a conventional antenna.

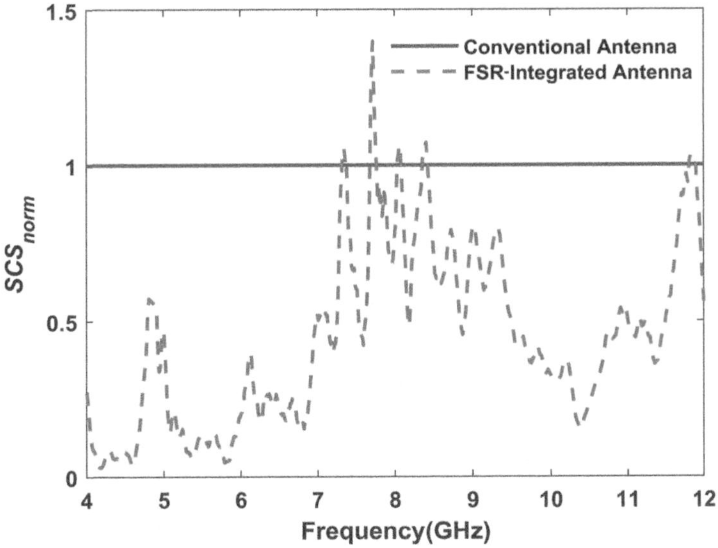

Figure 6.21 Normalized total bistatic RCS of the FSR-integrated antenna with reference to the conventional antenna [1].

Table 6.2 Performance Comparison of the Proposed FSR-Integrated Antenna with the Reported Radome-Type Low-RCS Antennas [1]

Ref.	*Operating Frequency*	*Size*	*Thick*	*No. of Layers*	*Lumped Element per Unit Cell*	*RCS Reduction Band*	*Max. Monostatic RCS Reduction*	*Average Monostatic RCS Reduction*	*Measured Average Total Bistatic RCS Reduction*
[16]	10 GHz	$1.8\lambda \times 1.8\lambda$	0.2λ	2	6	5.1–9.5 GHz and 13.2–16.2 GHz	20 dB	–	–
[17]	5 GHz	$1.8\lambda \times 1.8\lambda$	0.225λ	4	3	–	–	11 dB	–
[18]	10 GHz	$2.2\lambda \times 2.2\lambda$	0.8λ	5	12	6–18 GHz	11.8 dB	–	–
[19]	7.5 GHz	$1.5\lambda \times 1.5\lambda$	0.8λ	4	8	5–5.33 GHz and 9–13.4 GHz	–	–	–
[20]	10 GHz	$10.6\lambda \times 10.6\lambda$	0.5λ	2	2	2.9–7.6 GHz and 13.6–17.9 GHz	20 dB	–	–
This Work	8.4 GHz	$2.10\lambda \times 2.10\lambda$	0.7λ	3	4	4–7.5 GHz and 9.2–10.8 GHz	23 dB at 5.1 GHz	11.92 dB (4–7.5 GHz)/5.04 dB (9.2–10.8 GHz)	77% (4–7.5 GHz)/60% (9.2–10.8 GHz)

Note: λ =wavelength at the operating frequency of the antenna.

6.4 CONCLUSION

This chapter introduces an innovative approach for designing an absorptive-transmissive-absorptive (A-T-A) frequency selective rasorber (FSR) utilizing a modified frequency (MF) resonator. The proposed structure demonstrates enhanced selectivity at the transmission band, with a quality factor of 16.25. A cross-slot MF resonator, positioned on the front side of the resistive layer, generates a transmission pole that aligns with the passband of a double cross-slot- (DCS-) based frequency selective surface, thereby forming a narrow transmission window centered at 8.4 GHz within a wide absorption range spanning 3.4–10.8 GHz. To gain insight into its operation, an equivalent circuit model (ECM) is developed and analyzed. A prototype consisting of a 17 × 17 array of unit cells is fabricated, and measurement results confirm the design's effectiveness. Further, a 5 × 5 unit cell array of the FSR is used as a superstrate for a patch antenna to achieve reduced radar cross-section (RCS). The resulting FSR-integrated antenna demonstrates an average monostatic RCS suppression of 11.92 dB and 5.04 dB across the 4–7.5 GHz and 9.2–10.8 GHz bands, respectively, without compromising key antenna characteristics. Moreover, normalized total RCS values of 0.23 and 0.40 are attained in the lower and upper bands, respectively, compared to a conventional antenna.

REFERENCES

1. M. M. Zargar et al., "Frequency-Selective Rasorber Based on High-Q Minkowski Fractal-Shaped Resonator for Realizing a Low Radar Cross-Section Radiating System," IEEE Transactions on Electromagnetic Compatibility, vol. 64, no. 5, pp. 1574–1584, Oct. 2022.
2. M. Qu, S. Sun, L. Deng and S. Li, "Design of a Frequency-Selective Rasorber Based on Notch Structure," IEEE Access, vol. 7, pp. 3704–3711, 2019.
3. Q. Guo, Z. Li, J. Su, L. Y. Yang and J. Song, "Dual-Polarization Absorptive/Transmissive Frequency Selective Surface Based on Tripole Elements," IEEE Antenn. Wirel. Propag. Lett., vol. 18, no. 5, pp. 961–965, May 2019.
4. X. Xiu, W. Che, W. Yang, Y. Han and Q. Xue, "Double-Polarized Dual Passband Absorptive Frequency-Selective Transmission Structure," IEEE Trans. Electromagn. Compat., vol. 62, no. 5, pp. 1951–1960, Oct. 2020.
5. H. Ye, J. Wei, L. Lin, F. Liu, L. Miao, S. Bie and J. Jiang, "A Frequency Selective Surface Rasorber Based on Four Functional Layers," IEEE Trans. Antenn. Propag., vol. 69, no. 5, pp. 2768–2778, May 2021.
6. S. C. Bakshi, D. Mitra and F. L. Teixeira, "Wide-Angle Broadband Rasorber for Switchable and Conformal Application," IEEE Trans. Microw. Theory Techn., vol. 69, no. 2, pp. 1205–1216, Feb. 2021.
7. A. Parameswaran, D. Kundu and H. S. Sonalikar, "A Dual-Polarized Wideband Frequency-Selective Rasorber With Low in-Band Insertion Loss and High Oblique Incidence Stability," IEEE Trans. Electromagn. Compat., vol. 63, no. 6, pp. 1820–1828, Dec. 2021.

8. Q. Yu et al., "Miniaturized Wide-Angle Rasorber With a Wide Interabsorption High Transparent Bandpass Based on Multiple 2.5-D Resonators," IEEE Antenn. Wirel. Propag. Lett., vol. 21, no. 2, pp. 416–420, Feb. 2022.
9. M. Guo, T. Guo, Q. Cheng, Y. Zheng and Y. Fu, "Frequency Selective Rasorber With Anisotropic Transmission Band," IEEE Antenn. Wirel. Propag. Lett., vol. 20, no. 2, pp. 155–159, Feb. 2021.
10. H. Ye, W. Dai, X. Chen, H. Zhang, S. Bie and J. Jiang, "High-Selectivity Frequency-Selective Rasorber Based on Low-Profile Bandpass Filter," IEEE Antenn. Wirel. Propag. Lett., vol. 20, no. 2, pp. 150–154, Feb. 2021.
11. J. Xia et al., "Design of a Wideband Absorption Frequency Selective Rasorber Based on Double Lossy Layers," IEEE Trans. Antenn. Propag., vol. 68, no. 7, pp. 5718–5723, Jul. 2020.
12. M. Guo et al., "Broadband Absorptive Frequency-Selective Rasorber Based on Multilayer Resistive Sheets Using Multilayer Resonator," IEEE Trans. Antenn. Propag., vol. 70, no. 3, pp. 2009–2022, Mar. 2022.
13. W. Wan et al., "Chiral Absorber-Based Frequency Selective Rasorber With Identical Filtering Characteristics for Distinct Polarizations," IEEE Trans. Antenn. Propag., vol. 70, no. 5, pp. 3506–3514, May 2022.
14. B. A. Munk, "On Radar Cross Section of Antennas in General," Finite Antenna Arrays and FSS, Wiley-IEEE Press, pp. 15–55, 2003.
15. P. Alitalo, A. E. Culhaoglu, A. V. Osipov et al., "Experimental Characterization of a Broadband Transmission-Line Cloak in Free Space," IEEE Trans. Antenn. Propag., vol. 60, no. 10, pp. 4963–4968, Oct. 2012.
16. Q. Chen, M. Guo, D. Sang, Z. Sun and Y. Fu, "RCS Reduction of Patch Array Antenna Using Anisotropic Resistive Metasurface," IEEE Antenn. Wirel. Propag. Lett., vol. 18, no. 6, pp. 1223–1227, Jun. 2019.
17. Y. Han, L. Zhu, Y. Bo, W. Che and B. Li, "Novel Low-RCS Circularly Polarized Antenna Arrays via Frequency-Selective Absorber," IEEE Trans. Antenn. Propag., vol. 68, no. 1, pp. 287–296, Jan. 2020.
18. Y. Liu, Y. Jia, W. Zhang and F. Li, "Wideband RCS Reduction of a Slot Array Antenna Using a Hybrid Metasurface," IEEE Trans. Antenn. Propag., vol. 68, no. 5, pp. 3644–3652, May 2020.
19. W. Yu, Y. Yu, W. Wang, X. H. Zhang and G. Q. Luo, "Low-RCS and Gain Enhanced Antenna Using Absorptive/Transmissive Frequency Selective Structure," IEEE Trans. Antenn. Propag., vol. 69, no. 11, pp. 7912–7917, Nov. 2021.
20. B. Zhang, L. Li, C. Jin, Q. Lv and R. Mittra, "Wideband Low RCS Antenna Based on Hybrid Absorptive-Diffusive Frequency Selective Reflector," IEEE Access, vol. 9, pp. 77863–77872, Jun. 2021.

Chapter 7

Dual Transmission Miniaturized A-T-A-T FSR

In the preceding chapter, an A-T-A-type FSR incorporating an MF resonator was introduced to enhance passband selectivity. Despite this, applications involving dual-band radiating systems particularly for RCS reduction and EMI mitigation necessitate the development of FSRs with dual transmission features. Given the growing adoption of dual-/multiband antennas in modern communication systems, the design of such rasorbers has become increasingly important. This chapter presents a systematic approach to developing a compact A-T-A-T FSR offering dual transmission bands along with strong angular stability. The design methodology involves strategic use of low-frequency resonators on the resistive layer to support size reduction and a detailed investigation of cross-loop resonators to generate two distinct transmission poles through the lossy layer.

7.1 PROPOSED A-T-A-T FSR

7.1.1 Unit Geometry

Figure 7.1 illustrates the layout of the proposed FSR. The structure consists of two distinct layers: a top resistive (lossy) layer and a bottom lossless bandpass layer. Both layers are fabricated on a 0.762 mm thick Rogers Kappa substrate (dielectric constant $\varepsilon_r = 4.3$, loss tangent $\tan\delta = 0.005$) and are separated by a 5 mm air gap, as shown in Figure 7.1(a). The upper resistive layer, depicted in Figure 7.1(b), includes a central metallic square ring connected to four side-mounted parallel strips through rectangular patches loaded with 100 Ω chip resistors. A cross-loop resonator is embedded within the square ring to aid in the absorption response. The lower layer, shown in Figure 7.1(c), features a cross-slot at the center with additional quarter cross-slot elements positioned at each corner. The precise geometrical parameters are $p = 10, l_1 = 4.6, l_2 = 2.5, s_1 = 0.2, w_1 = 0.2, w_2 = 0.2, l_3 = 4.5, w_3 = 1, g = 0.2,$ $l_{p1} = 8.5, w_{p1} = 0.6, l_{p2} = 6.1, w_{p2} = 0.6, t = 0.762, H = 5$ (all dimensions in mm).

DOI: 10.1201/9781003644323-7

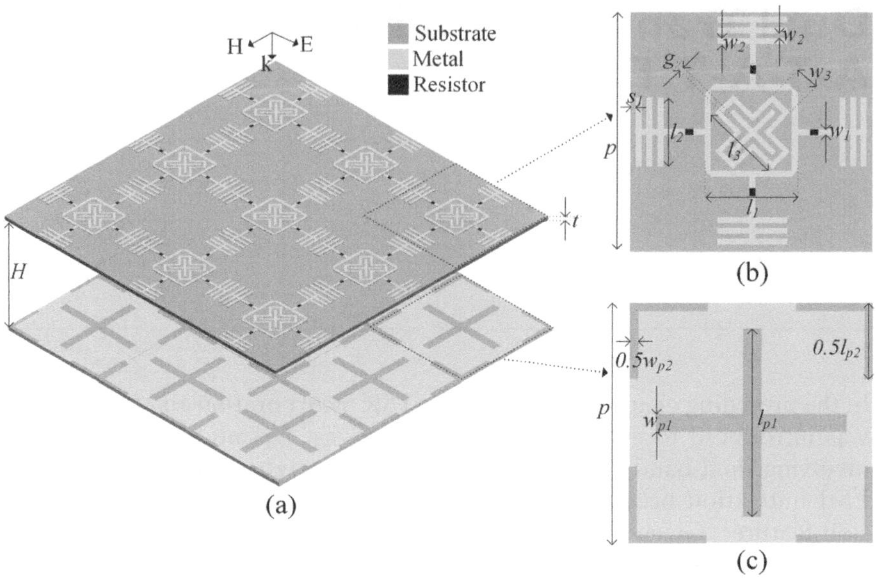

Figure 7.1 Geometry of the proposed FSR: (a) perspective view of 3 × 3 array, (b) top resistive, and (c) bottom bandpass layers [1].

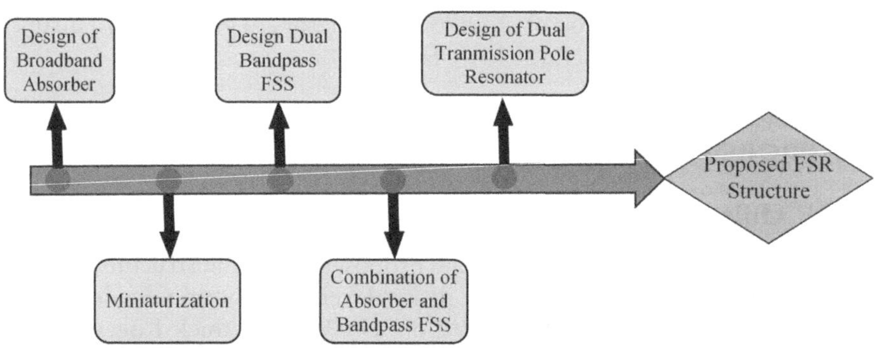

Figure 7.2 Design flow of the proposed FSR [1].

7.1.2 Design Process

The development of the proposed FSR follows a step-by-step approach, as illustrated in Figure 7.2. The design progression of the A-T-A-T-structured FSR begins with the realization of a broadband absorber and is elaborated in the following stages.

7.1.2.1 Broadband Absorber

A two-layer configuration is used to analyze the broadband absorber, where a resistive layer is positioned above a continuous metallic ground plane,

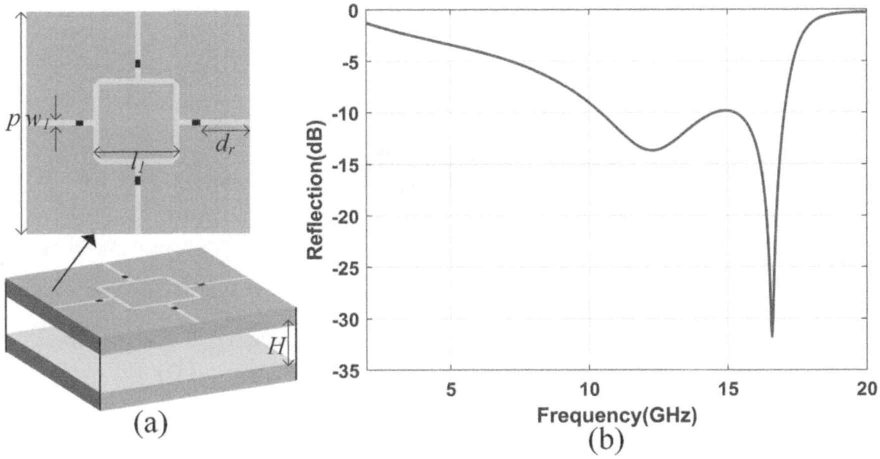

Figure 7.3 Broadband absorber: (a) unit cell schematic and (b) reflection coefficient [1].

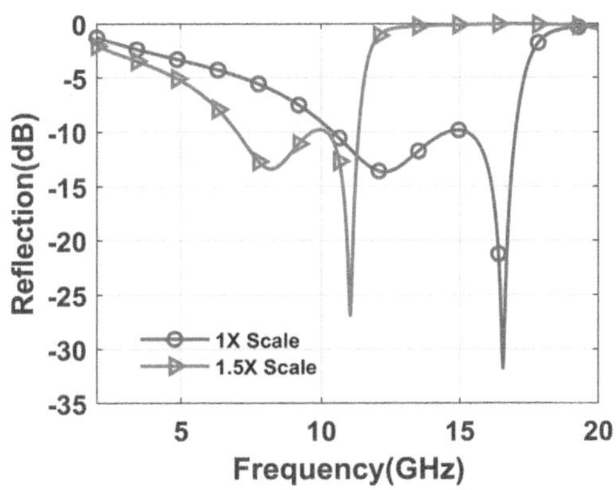

Figure 7.4 Reflection coefficient of broadband absorber with scaling effects [1].

separated by an air gap, as depicted in Figure 7.3(a). The resistive surface features a centrally located square ring, with rectangular arms on each side incorporating 100 Ω chip resistors. The structure is simulated using CST Studio Suite, and a wide 90% absorption bandwidth is achieved across the frequency range of 10.7–17.0 GHz, as shown in Figure 7.3(b). The gap (*H*) between the resistive and ground layers is 5 mm, which is approximately a quarter wavelength at the center frequency of the absorption band. The influence of unit cell dimensions on the absorption response is illustrated in Figure 7.4. With 1.5× scaling, the absorption shifts to a lower band (6.9–11.3 GHz), whereas the original (1×) size results in a higher frequency range (10.4–17.0 GHz). This indicates that simple dimensional scaling alters the operating frequency without reducing the electrical size, hence

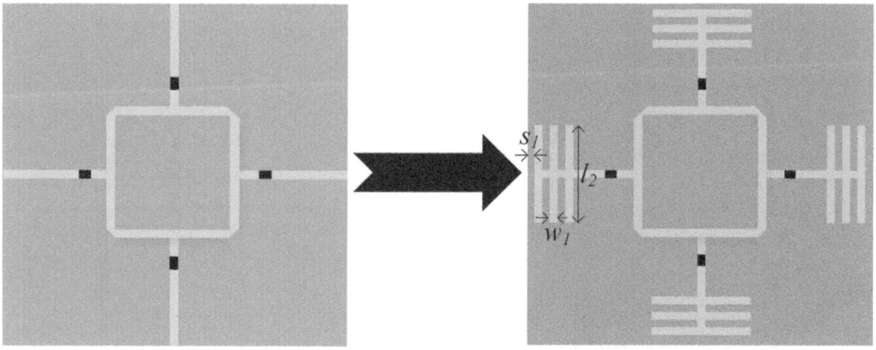

(a) (b)

Figure 7.5 Resistive layer of broadband absorber: (a) conventional and (b) modified [1].

failing to achieve effective miniaturization. The optimum parameters are $p = 10$, $l_1 = 4.6$, $w_1 = 0.2$, $d_r = 1.4$ (all dimensions in mm).

To reduce the electrical size of the unit cell, the absorber structure is revised by introducing additional low-frequency resonators, as depicted in Figure 7.5. The resistive surface shown in Figure 7.5(a) is modified by introducing a spacing of s_1 and incorporating three orthogonal resonators of length l_2 and width w_1 along each side of the rectangular strips, as illustrated in Figure 7.5(b). These resonators are intended to induce resonance at lower frequencies, thereby extending the absorption bandwidth in the lower frequency region. The impact of these added resonators is evident from the reflection coefficient plots in Figure 7.6, where a clear enhancement in low-frequency absorption is observed for the modified structure. A performance summary comparing both configurations is presented in Table 7.1. It is evident that with the inclusion of low-frequency resonators, the absorber bandwidth increases from 48.17% to 113.84%, accompanied by a reduction in electrical footprint (from 0.119 λ^2 to 0.019 λ^2) and thickness (from 0.22 λ to 0.09 λ). The optimum parameters are: $l_2 = 2.5$, $s_1 = w_1 = 0.2$ (all dimensions in mm).

7.1.2.2 Bandpass FSS

In developing the dual transmission FSR, a suitable dual-band bandpass FSS must be implemented, with resonances aligned to the absorption band. Initially, a central cross-slot structure with length l_{p1} and width w_{p1} is introduced at the center of the unit cell, resulting in a transmission band around 11.4 GHz (Figure 7.7[a]). This unit cell is dimensionally consistent with that of the broadband absorber. To create a second passband, quarter cross-slots with dimensions l_{p2} and w_{p2} are added at each corner of the unit cell, producing a second transmission window near 15.8 GHz,

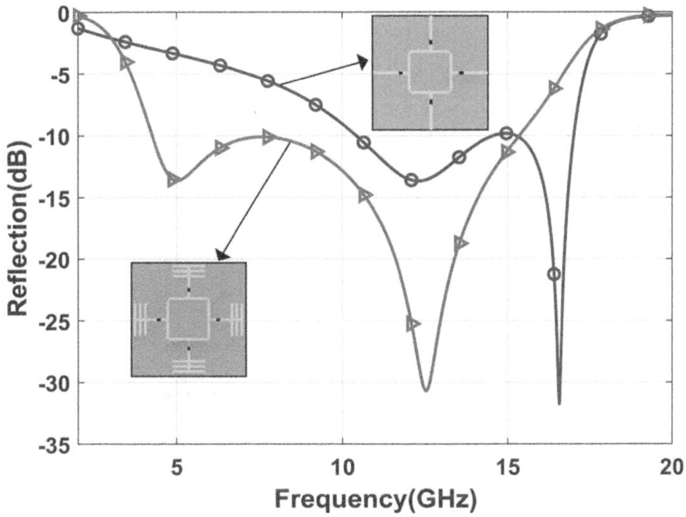

Figure 7.6 Reflection coefficient of the modified and conventional broadband absorber [1].

Table 7.1 Effect in Performance Parameters of Broadband Absorber due to Modification of Resistive Layer [1]

Absorber	*Bandwidth*	*Electrical Size*	*Electrical Thickness*
Basic	10.4–17.0 (GHz)	$0.119\lambda^2$	0.22λ
Modified	4.2–15.3 (GHz)	$0.019\lambda^2$	0.09λ

Note: λ is the wavelength at lowest frequency of absorption.

as shown in Figure 7.7(b). The impact of these slot features on the resonance frequencies is analyzed through parametric variation (Figure 7.8). Adjusting l_{p1} from 7.5 mm to 9.5 mm shifts the lower band from 12.9 GHz down to 9.9 GHz (Figure 7[a]), while the upper band remains fixed. In contrast, varying l_{p2} from 5.1 mm to 7.1 mm alters the upper resonance from 18.4 GHz to 13.9 GHz, keeping the lower band unchanged (Figure 7.8[b]). This dual-band FSS structure also behaves as a reflector in frequency regions outside the passbands. The optimum bandpass parameters are: $p = 10$, $l_{p1} = 8.5$, $w_{p1} = 0.6$, $l_{p2} = 6.1$, $w_{p2} = 0$. (all dimensions in mm).

7.1.2.3 Rasorber Design

In the subsequent step of the rasorber design, the dual-band bandpass FSS is integrated directly with the miniaturized absorber by substituting its metallic ground layer with the FSS structure. This configuration allows the FSS to serve as a reflective ground outside its passbands. The simulated performance of this

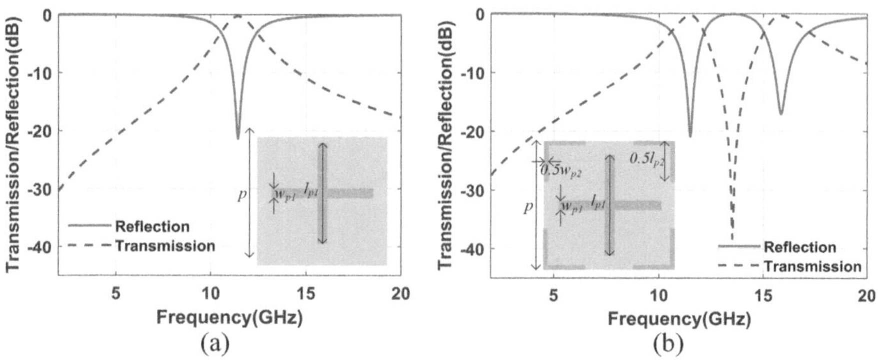

Figure 7.7 Bandpass FSS response: (a) single center slot and (b) with additional corner slots [1].

integrated setup is shown in Figure 7.9. It is evident that the resistive layer continues to deliver absorption outside the transmission regions. However, significant insertion losses (approximately 4 dB) are observed at the two passband frequencies. These losses are mainly due to the resistive layer overlapping with the transmission bands, resulting in energy dissipation where transparency is required. To resolve this issue, the resistive layer is redesigned to introduce two distinct transmission poles that align with the FSS operating frequencies, while still ensuring sufficient attenuation in the adjacent absorption ranges.

To independently generate two transmission poles within the wide absorption band, the resistive layer is examined in isolation. This is achieved by integrating a cross-loop resonator into the central square ring structure, as depicted in Figure 7.10. The embedded cross-loop element, characterized by parameters l_3, w_3, and a slot gap g, enables the formation of two transmission windows at 11.7 and 15.8 GHz, as shown in Figures 7.10(a) and 7.10(b). These resonances can be individually controlled by adjusting l_3 and g. As shown in Figure 7.10(c), modifying l_3 results in a concurrent shift of both transmission poles. Initially, the lower resonance can be set by varying l_3. Then, as illustrated in Figure 7.10(d), tuning the slot gap g allows for independent adjustment of the upper transmission frequency, while the lower resonance remains largely unaffected. This two-step tuning approach provides precise control over both passbands. The optimum parameter are: $l_3 = 4.5$, $w_3 = 1$, $g = 0.2$ (all dimensions in mm).

The final configuration is formed by integrating the resistive layer containing the cross-loop resonator with the dual-band bandpass FSS, and its performance is evaluated through simulation. As shown in Figure 7.12, the design exhibits distinct A-T-A-T behavior. The top resistive surface generates two wide absorption bands: one spanning 3.7–10.5 GHz with a bandwidth of 95.7%, and another spanning 12.6–14.6 GHz covering 14.70%. The combination of the cross-loop resonator and the underlying bandpass FSS

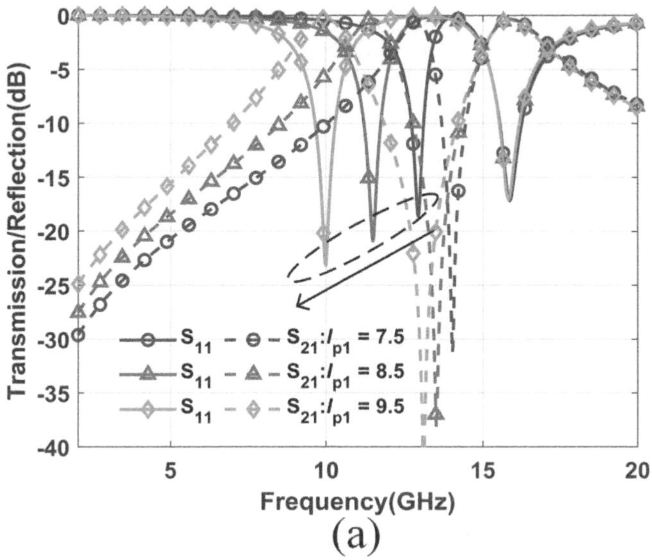

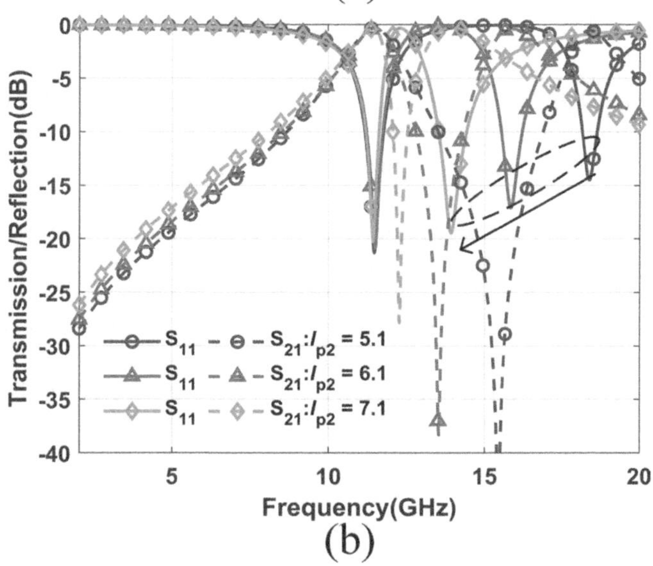

Figure 7.8 Variation in the response of bandpass FSS with variation in length of (a) center slot (lp_1) and (b) corner slot (lp_2) in mm [1].

produces two well-defined transmission bands centered at 11.7 GHz and 16.0 GHz, each with minimal insertion losses of 1.0 dB and 0.79 dB, respectively. The corresponding 3 dB bandwidths are 11.1–12.1 GHz (8.62%) and 15.3–16.7 GHz (8.75%). To examine the polarization response, the incident polarization angle is varied. Due to its four-fold symmetrical structure, the

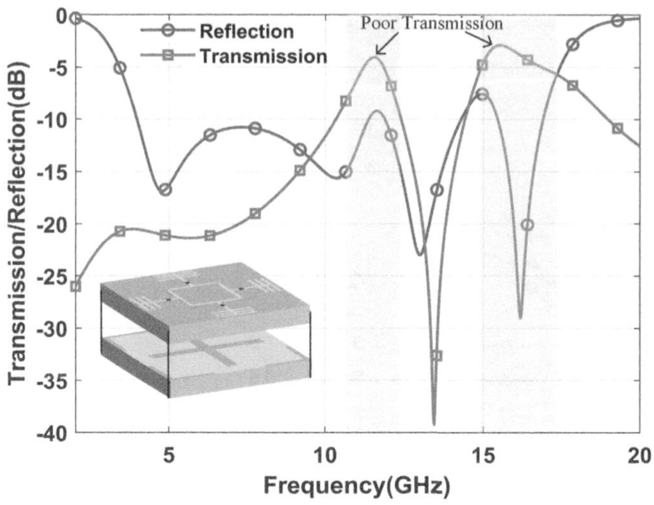

Figure 7.9 Transmission and reflection coefficients of the resistive layer in combination with bandpass FSS layer [1].

proposed rasorber maintains consistent behavior regardless of polarization, as illustrated in Figure 7.13.

The angular stability of the proposed FSR is evaluated for oblique incidences up to 60°. Separate plots for transmission and absorption characteristics under varying incidence angles are presented in Figure 7.14. The performance changes observed with increasing incidence angle are summarized in Table 7.2. Results indicate that both absorption and transmission remain relatively stable up to 50°, with only minor deviations. Specifically, at 50°, the absorption around 8 GHz slightly decreases from 80% to approximately 75%, though the bandwidth remains largely unaffected. Beyond this, at 60°, the absorption level falls below 80% across the band. Despite this reduction, the rasorber maintains effective angular stability for incidence angles up to 50°

7.1.3 Equivalent Circuit Analysis

To better understand the operational behavior of the proposed rasorber, an ECM is developed and shown in Figure 7.15. The ECM consists of two sequentially connected networks that represent the front resistive layer and the rear bandpass layer. The dielectric substrates are modeled using transmission line segments characterized by impedance $Z_d = Z_0 / \sqrt{\epsilon_r}$ and thickness t. The air gap between the two layers is represented as a transmission line section with characteristic impedance Z_0 and height H. The resulting network acts as a two-port system designed to match the free-space impedance at both input and output.

The circuit representation of the front resistive surface is derived by analyzing the layout and resonating components of the structure. Figure 7.16

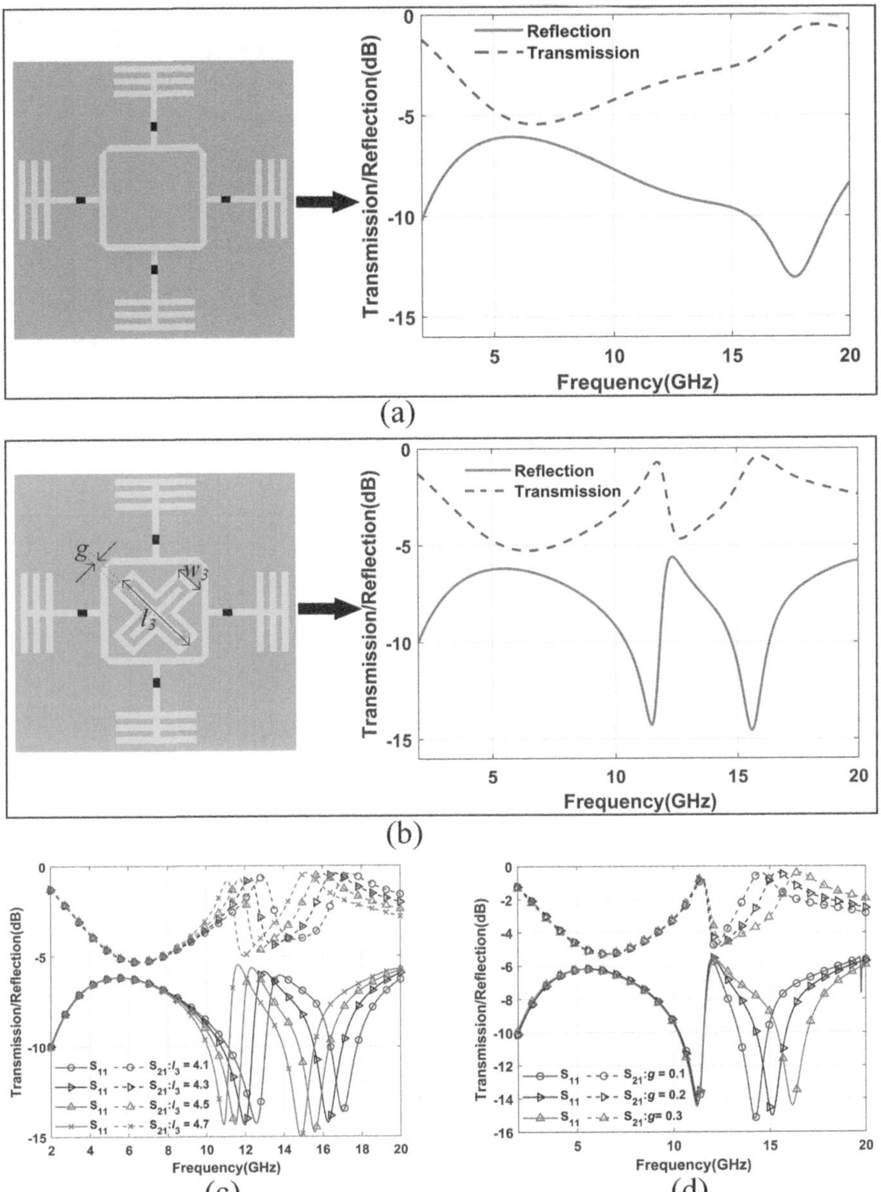

Figure 7.10 Transmission and reflection response of top resistive layer (a) without and (b) with cross-loop resonator; (c) with varying l_3 and (d) with varying g [1].

illustrates the corresponding schematic for the dominant polarization. The circuit paths associated with the central square ring and the peripheral rectangular arms incorporating low-frequency resonators are straightforward to identify. The equivalent model for the embedded cross-loop resonator

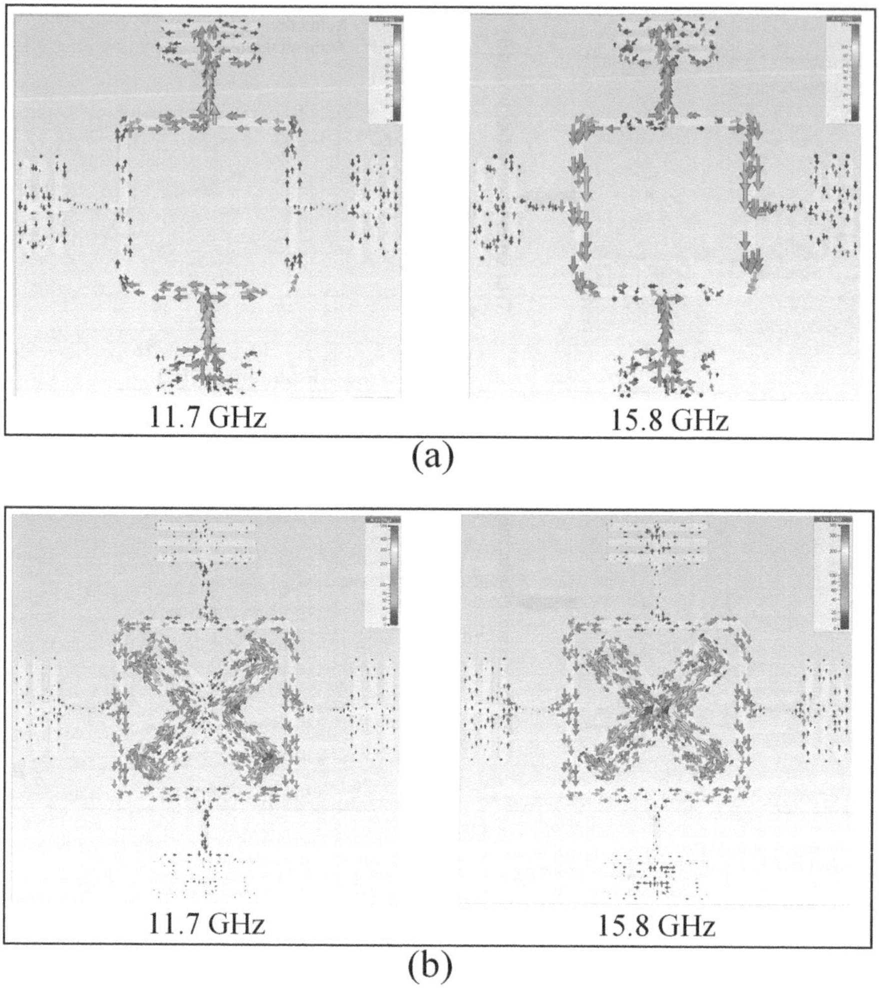

Figure 7.11 Surface currents on the resistive layers (a) without and (b) with cross-loop resonator [1].

is developed by observing the induced surface currents, as depicted in Figure 7.17. These currents show orthogonal flow for x- and y-polarized incidence. Under y-polarization, the current distribution leads to a central configuration of C_2 and L_4, flanked by two identical branches consisting of $L_2^{'}, L_3^{'}$ and $C_3^{'}$. These are represented in the ECM as a simplified circuit with parallel $L_3 - C_3$ and series L_2. The dual bandpass response of the rear FSS layer, resulting from the center and corner cross-slots, is modeled using two parallel resonant branches: $L_{P1} - C_{P1}$ and $L_{P2} - C_{P2}$.

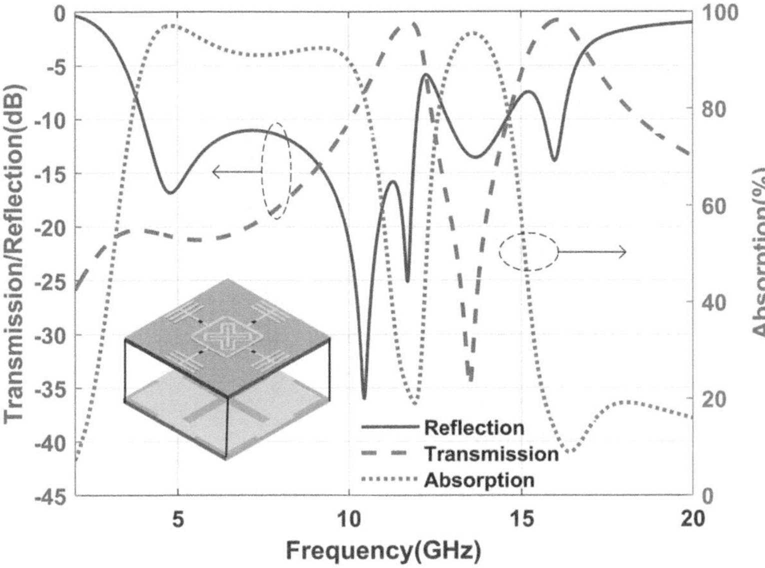

Figure 7.12 Simulated transmission, reflection, and absorption of the proposed A-T-A-T FSR [1].

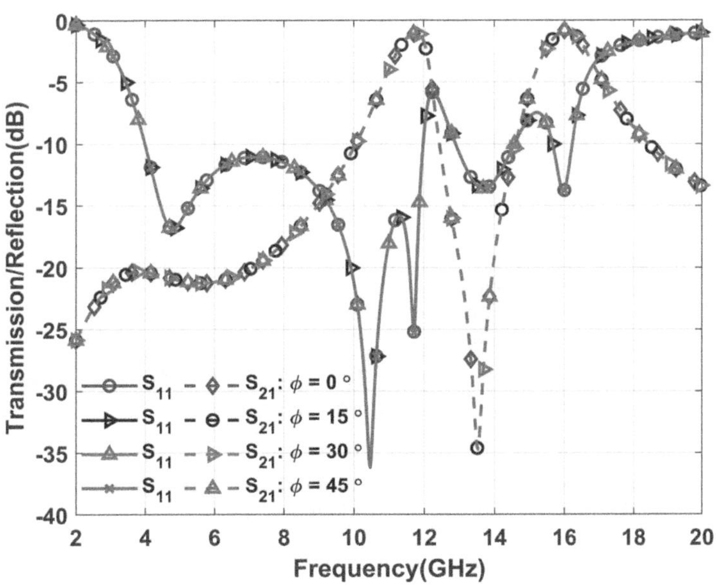

Figure 7.13 Response of the proposed FSR with polarization variations [1].

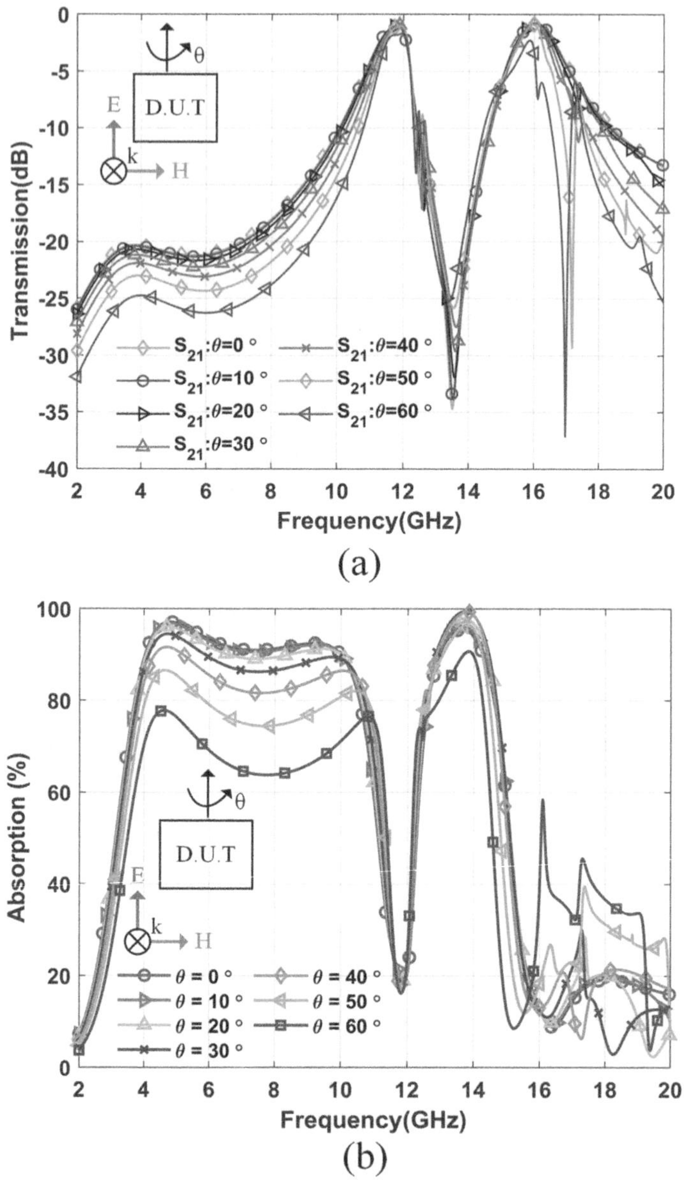

Figure 7.14 (a) Transmission coefficient and (b) absorption plots of the proposed FSR under various oblique incidence [1].

The reflection (S_{11}) and transmission (S_{21}) coefficients are computed from the overall ABCD matrix of the ECM using standard S-parameter conversion formulas.

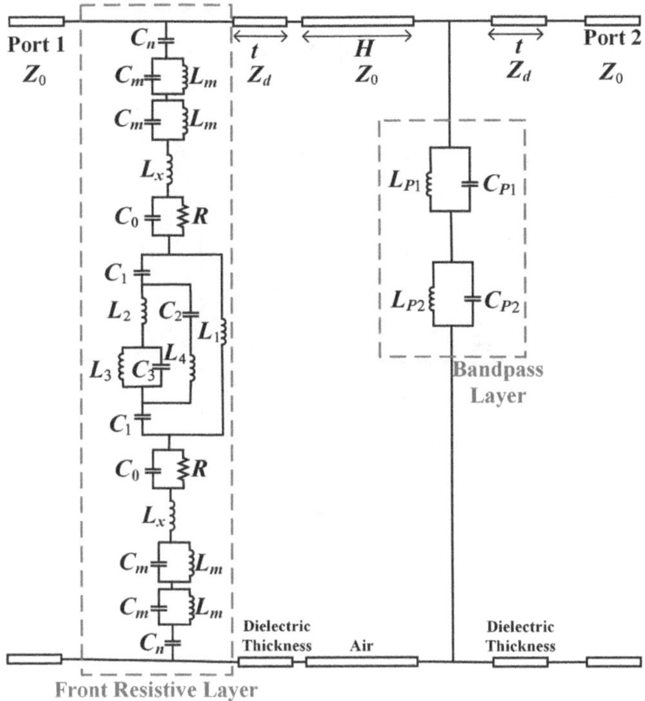

Figure 7.15 ECM of the proposed A-T-A-T FSR [1].

Table 7.2 Performance of the Proposed FSR under Oblique Incidence [1]

Angle	*A1 (GHz)*	*T1 (GHz)*	*A2 (GHz)*	*T2 (GHz)*
0°	3.7–10.5	11.1–12.1 (1.0 dB)	12.6–14.6	15.3–16.7 (0.7 dB)
10°	3.7–10.5	11.1–12.1 (1.0 dB)	12.6–14.6	15.3–16.7 (0.7 dB)
20°	3.7–10.6	11.1–12.1 (1.0 dB)	12.5–14.6	15.3–16.6 (0.7 dB)
30°	3.7–10.6	11.1–12.1 (0.9 dB)	12.5–14.6	15.3–16.5 (0.7 dB)
40°	3.8–10.7	11.2–12.1 (0.8 dB)	12.4–14.6	15.4–16.4 (0.7 dB)
50°	4.0–5.8 and 9.8–10.8	11.3–12.1 (1.0 dB)	12.6–14.4	15.4–16.2 (0.8 dB)
60°	Below 80%	11.4–12.1 (1.7 dB)	12.9–14.2	15.6–15.9 (2.2 dB)

Note: A1/A2 is the lower/upper absorption band; T1/T2is the lower/upper transmission band with minimum insertion loss.

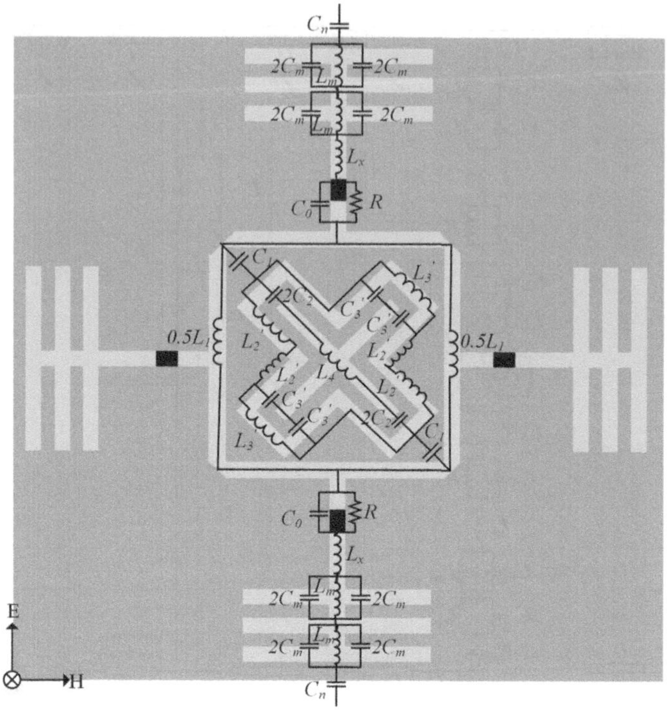

Figure 7.16 Circuit schematic on the resistive layer [1].

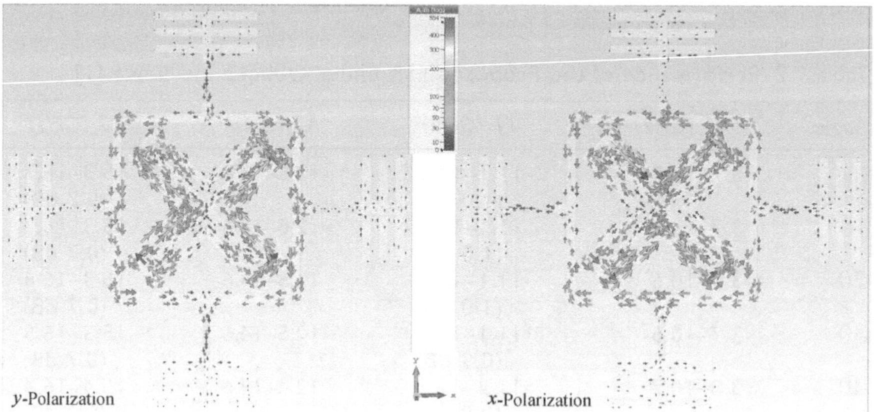

Figure 7.17 Current distribution on the resistive layer under y- and x-polarizations [1].

$$S_{11} = \frac{A + B / Z_0 - CZ_0 - D}{A + B / Z_0 + CZ_0 + D} \tag{7.1}$$

$$S_{21} = \frac{2}{A + B / Z_0 + CZ_0 + D} \tag{7.2}$$

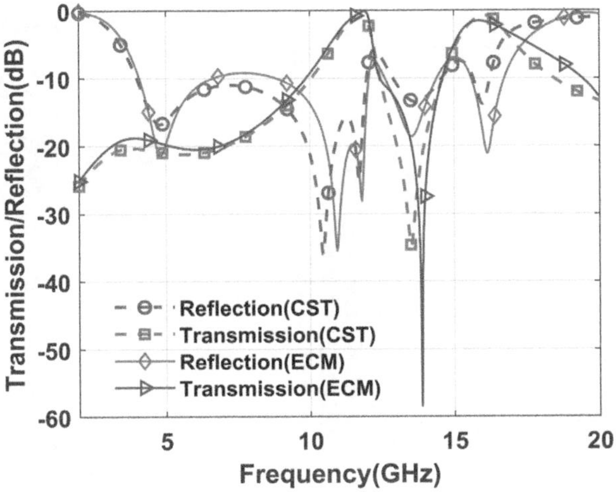

Figure 7.18 Transmission and reflection coefficient of the ECM in comparison with full-wave EM simulated response of proposed FSR [1].

The complete ABCD matrix is derived by sequentially multiplying the individual ABCD matrices of the cascaded sections:

$$\begin{bmatrix} A & B \\ C & D \end{bmatrix} = \begin{bmatrix} 1 & 0 \\ 1/Z_F & 1 \end{bmatrix} \times \begin{bmatrix} \cos\beta t & jZ_d \sin\beta t \\ j\sin\beta t / Z_d & \cos\beta t \end{bmatrix} \times \begin{bmatrix} \cos\beta_0 H & jZ_0 \sin\beta_0 H \\ j\sin\beta_0 H / Z_0 & \cos\beta_0 H \end{bmatrix} \times \begin{bmatrix} \cos\beta t & jZ_d \sin\beta t \\ j\sin\beta t / Z_d & \cos\beta t \end{bmatrix} \times \begin{bmatrix} 1 & 0 \\ 1/Z_P & 1 \end{bmatrix} \tag{7.3}$$

Here, the propagation constant β is given by $\beta = \beta_0\sqrt{\mu_r\epsilon_r}$, where μ_r and ϵ_r are the substrate's relative permeability and permittivity. The parameters Z_F and Z_P represent the impedances of the front resistive surface and the rear bandpass layer, respectively.

The complete network is evaluated using the Keysight ADS circuit simulation tool. The starting estimates for the circuit component values, based on the corresponding physical dimensions, are derived using standard equations available in the literature [2].

$$C = \epsilon_0\epsilon_{eff}\frac{2p}{\pi}\log\left(\frac{1}{\sin\frac{\pi s}{2p}}\right) \tag{7.4}$$

$$L = \mu_0\frac{p}{2\pi}\log\left(\frac{1}{\sin\frac{\pi W}{2p}}\right) \tag{7.5}$$

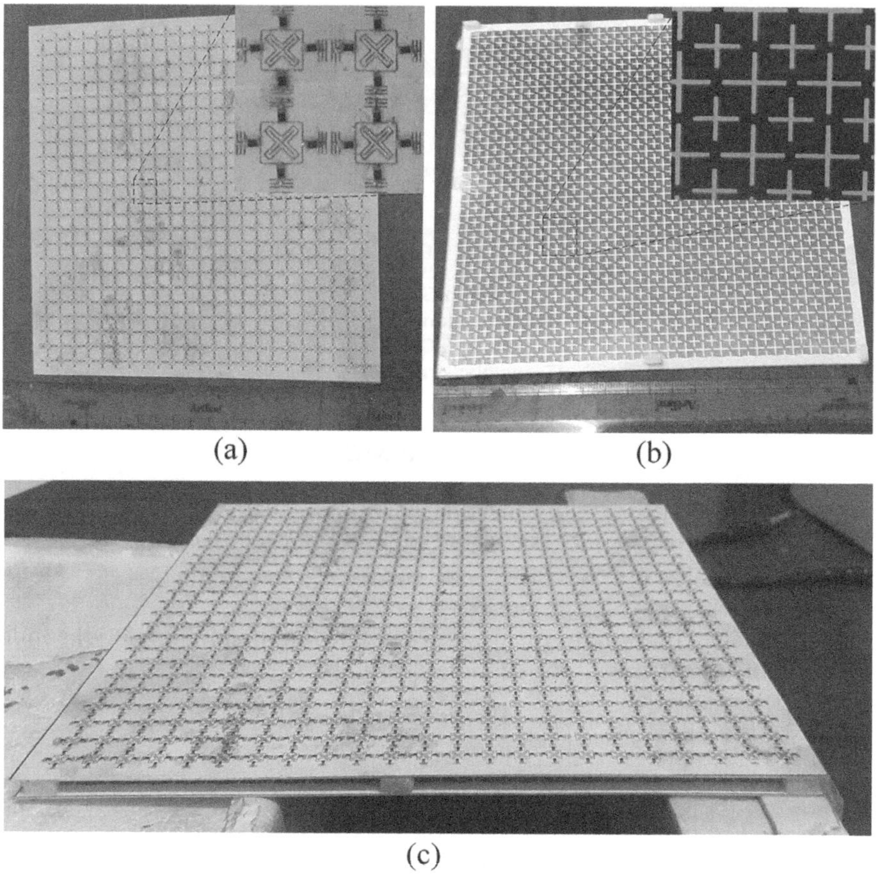

Figure 7.19 Fabricated A-T-A-T rasorber prototype: (a) top, (b) bottom, and (c) perspective view [1].

In this context, p denotes the unit cell size, s is the spacing between adjacent patches, and w is the width of the metal strip. ϵ_0 and μ_0 represent the permittivity and permeability of free space, respectively, while ϵ_{eff} is the substrate's effective permittivity. Initial estimates for the inductance and capacitance values are derived from these physical parameters and later refined using parameter sweep and optimization features in the circuit simulator. The final optimized values for each circuit element are: $C_n = 0.22097\text{pF}, C_m = 7.50025\text{pF}, L_m = 1.15077\text{nH}, L_x = 2.07095\text{nH}, R = 100\Omega, C_0 = 0.001\text{pF}, C_1 = 0.34289\text{pF}, L_1 = 0.74098\text{nH}, L_2 = 0.40096\text{nH}, C_2 = 0.40092\text{pF}, L_3 = 17.01083\text{nH}, C_3 = 0.803012\text{pF}, L_4 = 0.001\text{nH}, L_{P1} = 0.3971\text{nH}, C_{P1} = 0.45091\text{pF}, L_{P2} = 0.3055\text{nH}, C_{P2} = 0.31097\text{pF}, Z_0 = 377\Omega, Z_d = 181\Omega, t = 0.8\text{mm}, H = 5\text{mm}$.

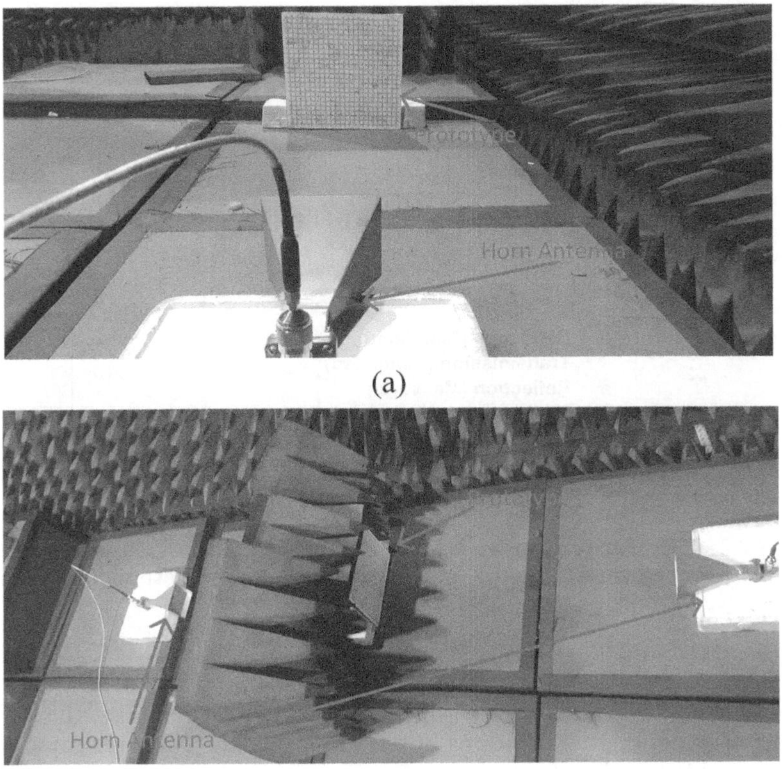

(a)

(b)

Figure 7.20 Experimental setup: (a) reflection and (b) transmission measurements [1].

A comparison of the reflection and transmission responses between the circuit model and full-wave electromagnetic simulations is shown in Figure 7.18, demonstrating strong agreement and confirming the validity of the proposed FSR design.

7.2 EXPERIMENTATION AND DISCUSSIONS

A prototype consisting of a 23 × 23 array of the proposed unit cell is fabricated, covering a total area of 230 mm × 230 mm. The structure is fabricated using a Rogers Kappa substrate characterized by a relative permittivity (ϵ_r) of 4.3, loss tangent ($\tan\delta$) of 0.005, and a thickness of 0.762 mm. Images of the fabricated top (resistive) and bottom (bandpass) layers are shown in Figures 7.19(a) and 7.19(b), respectively. To implement the resistive

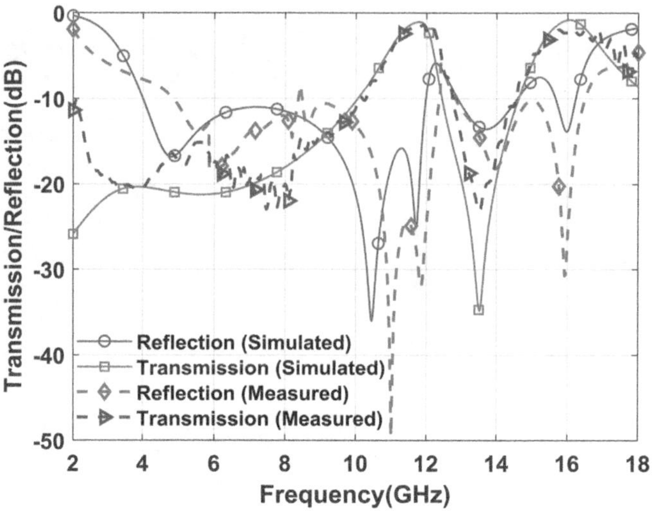

Figure 7.21 Comparison between the measured transmission and reflection coefficients of the prototype and the corresponding simulated response [1].

functionality, 100 Ω chip resistors (VISHAY CRCW0603100RFKEA) are soldered at designated positions in the top layer. The two fabricated layers are assembled with a 5 mm air gap using plastic spacers, as illustrated in Figure 7.19(c). Performance evaluation is conducted in an anechoic chamber using a free-space setup, where standard gain horn antennas covering S, C, J, X, and Ku bands are used along with a Keysight PNA N5224B vector network analyzer for measuring reflection and transmission characteristics.

For reflection measurements, a single horn antenna is employed, and the reflection coefficient is obtained using the time-gating function of the vector network analyzer. The experimental setup for reflection is illustrated in Figure 7.20(a). A metallic copper plate of the same size is used to acquire the reference reflection, and the prototype's reflection data are normalized against this baseline. In the case of transmission measurements, the arrangement shown in Figure 7.20(b) is used, where the prototype is positioned between two horn antennas. The reference for transmission is taken from the direct signal path between the antennas without the sample. The comparison between simulated and measured transmission and reflection results is presented in Figure 7.21. A strong correlation is observed, confirming the validity of the design. Minor discrepancies in the measured results are likely due to slight imperfections during the soldering of the chip resistors.

The fabricated prototype is evaluated for performance under oblique incidence angles up to 40°. The complete measurement setup for evaluating reflection and transmission at inclined angles is described in detail in earlier

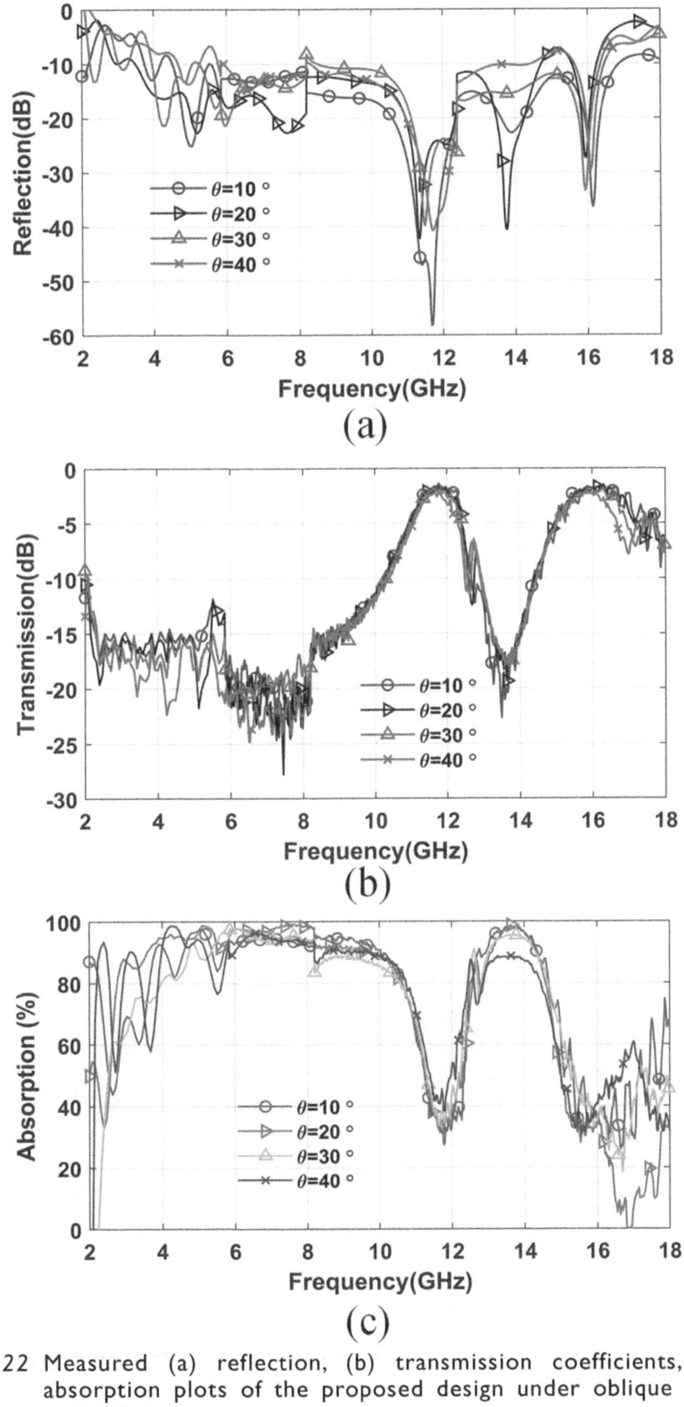

Figure 7.22 Measured (a) reflection, (b) transmission coefficients, and (c) absorption plots of the proposed design under oblique incidence angles [1].

Table 7.3 Performance Comparison of the Proposed FSR with the Previous Reported A-T-A-T Rasorbers [1]

Ref.	*Operating Band (GHz)*	*Passbands (I.L.)*	*Lower Absorption (GHz)*	*Upper Absorption (GHz)*	*Thickness*	*Unit Cell Size*	*Polarization*	*Angular Stability*
[3]	4.02–13.05 (105.79%)	7.2 GHz (2.3 dB), 13.05 GHz (1.69 GHz)	4.02–6.27 (43.73%)	8.21–12.02 (37.66%)	8 mm (0.107λ)	9 mm x 18 mm (0.028λ^2)	Single	20°
[4]	3.3–10.1 (101.49%)	6.1 GHz (0.06 dB)/10.1 GHz 0.2 dB)	3.3–4.97 (40.38%)	7.42–9.03 (19.57%)	9 mm (0.09λ)	8 mm hexagon (0.020λ^2)	Dual	30°
[5]	4.5–11.9 (90.24%)	8 GHz (0.39 dB), 11.9 GHz (0.64 dB)	4.5–7.3 (47.45%)	8.5–11.0 (25.64%)	7 mm (0.105λ)	15 mm x 15 mm (0.05λ^2)	Dual	30°
This work	3.7–16.0 (124.87%)	11.7 GHz (1.0 dB)/16.0 GHz (0.7 dB)	3.7–10.5 (95.7%)	12.6–14.6 (14.70%)	6.52 mm (0.08λ)	10 mm x 10 mm (0.015λ^2)	Insensitive	50°

Note: λ corresponds to the lowest frequency of absorption.

chapters. The experimental results for reflection and transmission at different incident angles up to 40° are shown in Figures 7.22(a) and 7.22(b), respectively. Corresponding absorption curves for each angle of incidence are displayed in Figure 7.22(c). The observed responses under angular incidence confirm satisfactory absorption and transmission performance, thereby experimentally demonstrating the angular robustness of the fabricated structure.

A performance comparison between the proposed rasorber and previously reported A-T-A-T FSR configurations [3–5] is summarized in Table 7.3. The advancements achieved in this work over earlier A-T-A-T designs can be outlined as follows:

1. The presented structure features a compact electrical footprint of $0.015\lambda^2$, which is smaller than those in [3–5].
2. It also demonstrates a reduced electrical thickness of 0.08λ, offering a thinner profile than prior designs [3–5].
3. The proposed FSR achieves a broader operational bandwidth of 124.87%, exceeding that of earlier reported counterparts [3–5].
4. A lower absorption band of 95.7% is realized, outperforming other A-T-A-T structures [3–5] in terms of absorption efficiency.
5. Compared to existing designs [3–5], the suggested rasorber provides enhanced angular stability, maintaining reliable performance up to 50° incidence.

7.3 CONCLUSION

This chapter presents a compact and low-profile dual-transmission A-T-A-T FSR. The design evolution is detailed in sequential stages, beginning with a broadband miniaturized absorber, followed by the integration of a dual-band bandpass FSS and culminating in the introduction of dual transmission poles via a cross-loop resonator embedded in the resistive layer. The final structure features a compact footprint of $0.015\lambda^2$ and a thickness of 0.08λ. It offers two transmission windows at 11.7 GHz (8.62%) and 16.0 GHz (8.75%), flanked by wide absorption bands spanning 3.7–10.5 GHz (95.7%) and 12.6–14.6 GHz (14.70%). The FSR also maintains stable performance up to 50° incidence angle. The operation mechanism is analyzed through surface current analysis and an equivalent circuit model. Experimental results from the fabricated prototype, tested under both normal and angled incidences, confirm the simulated performance.

REFERENCES

1. M. M. Zargar et al., "Miniaturized Design of Dual Transmission Frequency Selective Rasorber With Wide Angular Stability," IEEE Open J. Antenn. Propag., vol. 5, no. 4, pp. 922–932, Aug. 2024.

2. H. B. Wang, Y. J. Cheng and Z. N. Chen, "Wideband and Wide-Angle Single-Layered-Substrate Linear-to-Circular Polarization Metasurface Converter," IEEE Trans. Antenn. Propag., vol. 68, no. 2, pp. 1186–1191, Feb. 2020.
3. M. Guo, Q. Chen, Z. Sun, D. Sang and Y. Fu, "Design of Dual-Band Frequency-Selective Rasorber," IEEE Antenn. Wirel. Propag. Lett., vol. 18, no. 5, pp. 841–845, May 2019.
4. X. Zhang, W. Wu, J. Huang, W. Zhang, Y. Ye and N. Yuan, "Dual Polarized Frequency Selective Rasorber With Two Transmission Bands," IEEE Access, vol. 7, pp. 139795–139801, 2019.
5. Y. Jia, A. Wu, Y. Liu, W. Zhang and Z. Zhou, "Dual-Polarization Frequency-Selective Rasorber With Independently Controlled Dual-Band Transmission Response," IEEE Antenn. Wirel. Propag. Lett., vol. 19, pp. 831–835, 2019.

Chapter 8

Low-RCS Crossed Dipole Antenna Co-Designed with AFSR Structure

Chapters 2 through 7 have presented diverse design methodologies for different types of frequency selective rasorbers (FSRs). These FSRs, also referred to as absorptive frequency selective transmission (AFST) structures, are characterized by the coexistence of transmission and absorption bands, making them suitable for integration as radomes to reduce the radar cross-section (RCS) of radiating elements. Alternatively, RCS suppression can also be achieved using band-rejecting absorbers, commonly termed absorptive frequency selective reflection (AFSR) structures, which serve effectively as reflective ground planes. This chapter focuses on exploring the integration of a dual-polarized crossed dipole antenna with a polarization-independent AFSR to achieve RCS reduction. Initially, a compact, polarization-insensitive AFSR unit is developed by embedding a bandstop ring resonator within a circular-cross-shaped broadband absorber. The dual-polarized antenna is then innovatively co-designed with the AFSR, ensuring the AFSR functions as a reflecting surface at the antenna's operating frequency. The resulting integrated system demonstrates significant RCS suppression under both TE and TM polarized incident waves compared to a conventional antenna backed by a metallic reflector.

8.1 AFSR STRUCTURE

8.1.1 Design and Analysis

Figure 8.1 illustrates the structural configuration of the proposed AFSR unit cell, which consists of a two-layer assembly: a resistive surface on the top and a metallic ground plane at the bottom, spaced apart by a height H (as shown in Figure 8.1(a)). The layout is fabricated on an FR-4 substrate with a thickness of 0.8 mm. The top resistive surface features a circular-cross-shaped conductive pattern, where four 200 Ω surface-mount chip resistors are placed across the rectangular strip gaps (Figure 8.1(b)). On the reverse side of this resistive layer, a circular ring resonator made of metal is printed. Electromagnetic analysis and simulation of the design are

DOI: 10.1201/9781003644323-8

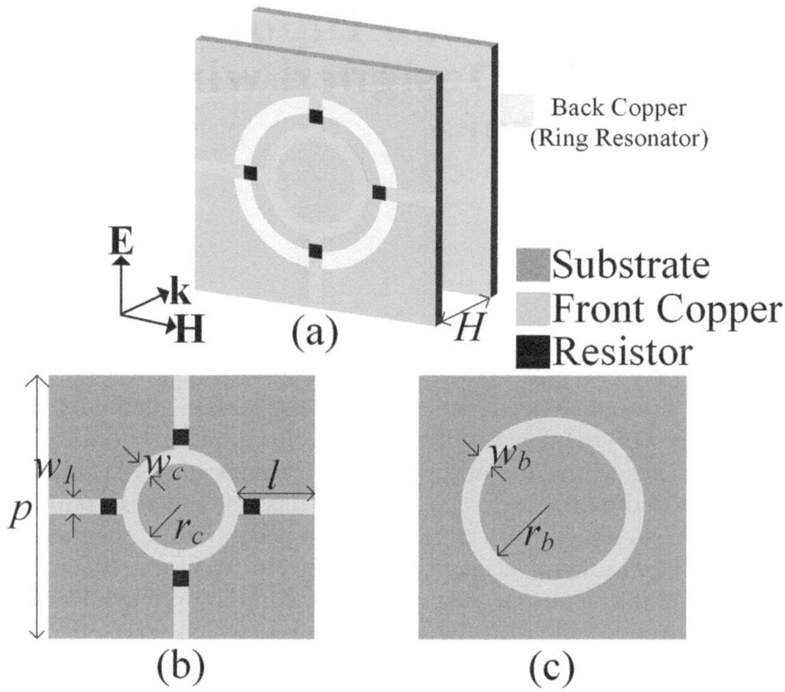

Figure 8.1 AFSR unit cell geometry: (a) perspective view, (b) front side, and (c) back side of top resistive layer [1].

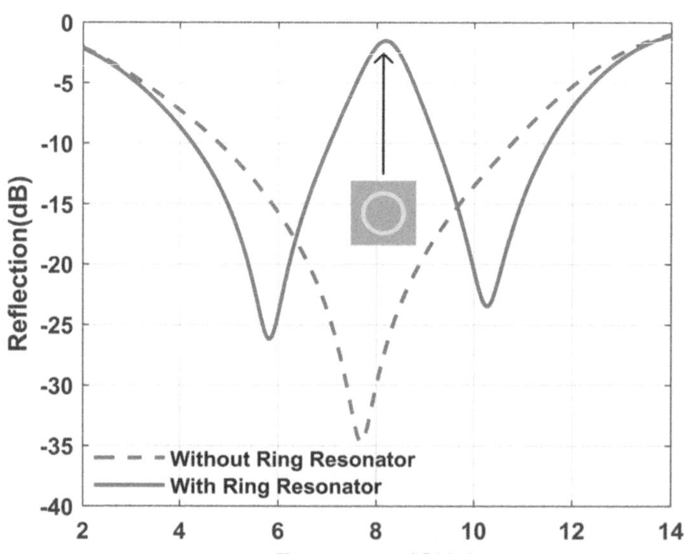

Figure 8.2 Simulated reflection coefficients of the AFSR structure and the broad-band absorber [1].

performed using CST Microwave Studio. The optimum dimensions (in mm) are: $p = 15$, $r_c = 2.2$, $w_c = 1$, $r_b = 3.2$, $w_1 = 0.8$, $w_b = 0.6$, $l = 4.3$, and $H = 8$.

Positioning the resistive surface which comprises of a circular-cross resonator loaded with chip resistors, at a quarter-wavelength distance ($\lambda/4$) from the ground plane facilitates wideband electromagnetic wave absorption, as illustrated in Figure 8.2. This configuration yields over 90% absorption within the 4.7–10.7 GHz frequency range. To introduce a reflective notch centered within this absorption range, a metallic circular ring is etched onto the rear face of the resistive layer. Acting as a bandstop frequency selective surface, the ring creates a reflection dip at approximately 8.2 GHz. As observed in Figure 8.2, the presence of the ring splits the original wide absorption region into two separate absorption bands, 4.2–7.0 GHz and 9.2–11.5 GHz, while a pronounced reflective notch appears at the targeted frequency.

To analyze the operational behavior of the proposed AFSR configuration, an equivalent circuit model (ECM) is developed and illustrated in Figure 8.3. This ECM represents a one-port network where the models for the resonant elements on both the front and rear surfaces of the resistive layer are connected in series. On the front side, the circular-cross structure is characterized by a resistor–capacitor pair ($R||C_0$) that accounts for the lumped resistance and gap capacitance, while the inductive response of the rectangular strip is captured by an inductor L_0. The circular ring's contribution is represented using a parallel L_1–C_1 circuit. On the rear side of the resistive layer, the circular ring is modeled with another parallel resonant circuit ($L_2||C_2$), and the mutual capacitance between structures is captured by C_3. The dielectric spacer is modeled as a transmission line with impedance $Z_d = Z_0 / \sqrt{\epsilon_r}$ and length equal to the substrate thickness t, where Z_0 is the free space impedance. The air gap between the resistive sheet and the ground plane is also represented by a transmission line with impedance Z_0 and length H. The optimum values of circuit parameters are: $R = 200\ \Omega$, $C_0 = 0.051$ pF, $L_0 = 2.45$ nH, $C_1 = 0.0011$ pF, $L_1 = 3.185$ nH, $C_2 = 0.1$ fF, $L_2 = 39.305$ nH, $C_3 = 0.0091$ pF, $Z_d = 181\ \Omega$, $t = 0.8$mm, $Z_0 = 377\ \Omega$.

The equivalent circuit model (ECM) of the proposed AFSR structure is evaluated using the Keysight ADS simulation tool. A comparison between the ECM-derived reflection coefficient and the full-wave simulation results obtained from CST is presented in Figure 8.4. Figure 8.4(a) illustrates the ECM response excluding the rear-side ring resonator, revealing a wideband absorption behavior that aligns closely with the CST-based results for the broadband absorber. In contrast, Figure 8.4(b) displays the ECM response for the complete AFSR structure, where the introduction of the circuit model for the back side ring produces a distinct reflection notch at the center frequency. The strong correlation between the ADS and CST simulations confirms the validity of the ECM and supports the functional understanding of the proposed AFSR configuration.

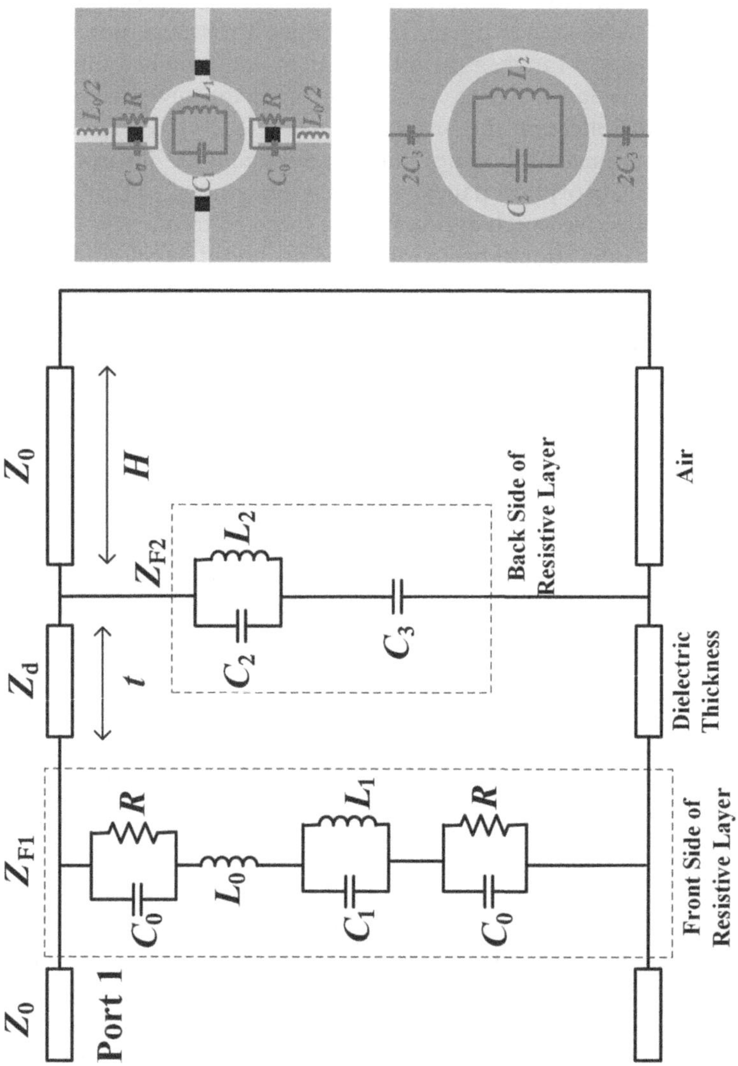

Figure 8.3 ECM of the proposed AFSR structure [1].

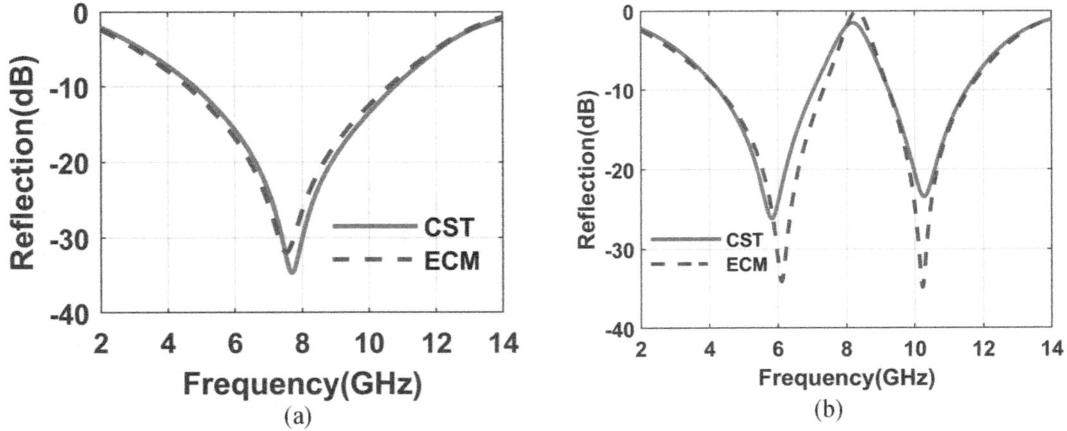

Figure 8.4 ECM response in comparison with the CST results: (a) broadband absorber and (b) AFSR structure [1].

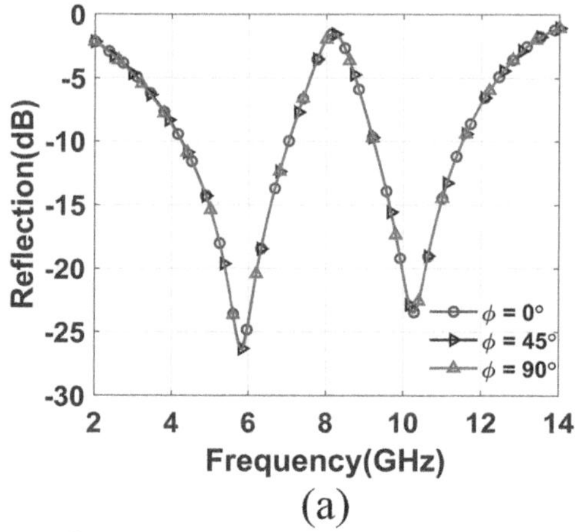

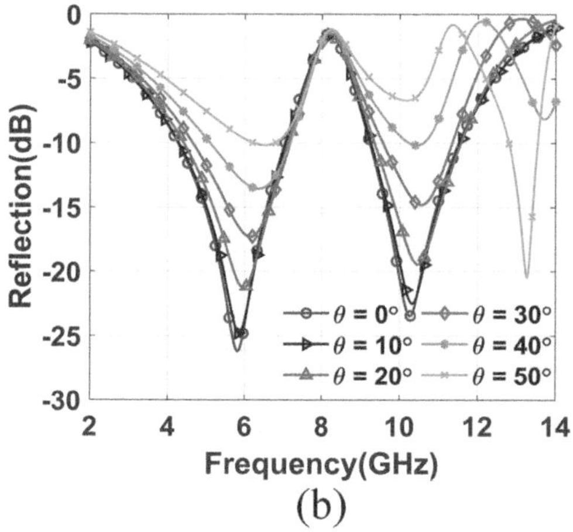

Figure 8.5 Simulated responses of the proposed AFSR structure with respect to different (a) polarizations and (b) incidence angles [1].

The electromagnetic response of the proposed AFSR structure is evaluated under different polarization orientations of the incoming wave, as illustrated in Figure 8.5(a). The stability of the reflection profile across these polarization variations confirms that the structure exhibits polarization-insensitive characteristics. Additionally, the performance of the AFSR is assessed for oblique incidences up to an angle of 50°. It is observed that the structure maintains stable absorption behavior up to 30°, beyond which the absorption bands begin to deteriorate, indicating a reduction in angular robustness at higher incidence angles.

8.1.2 Fabrication and Measurements

To experimentally verify the AFSR design, a prototype comprising a 21 × 21 array of the proposed unit cells is developed on an FR-4 substrate

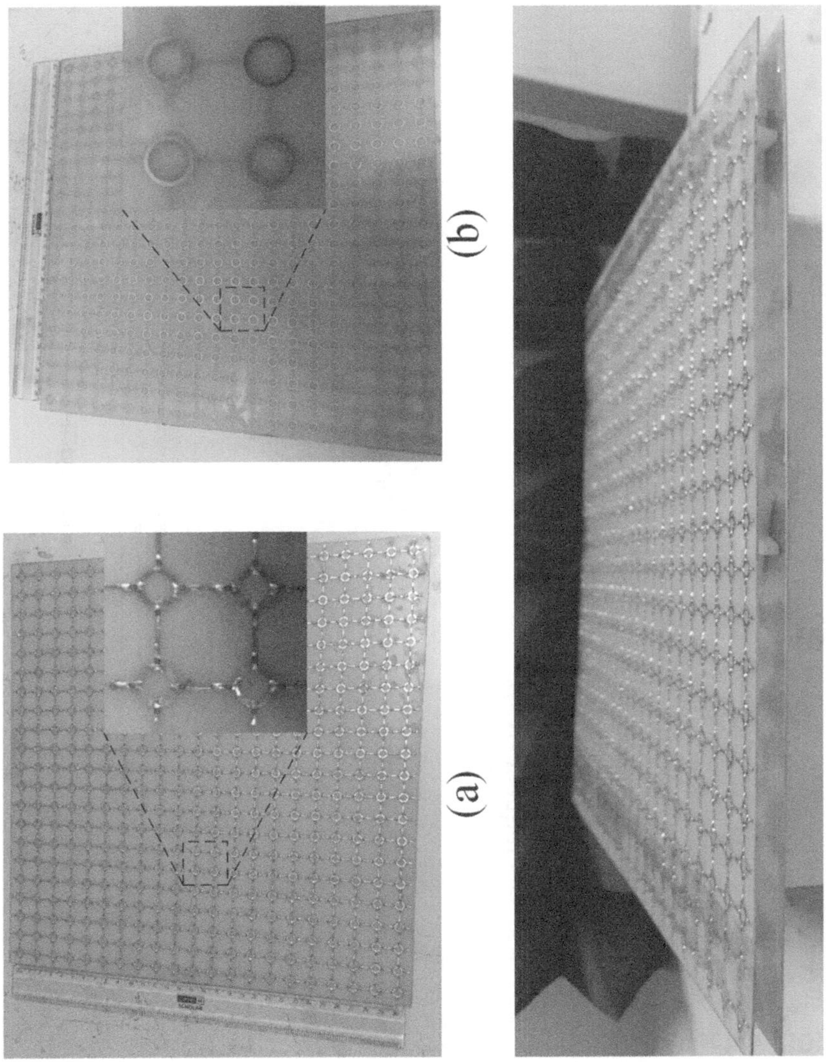

Figure 8.6 Photograph of the fabricated AFSR prototype: (a) front and (b) back side of the top layer; (c) side view [1].

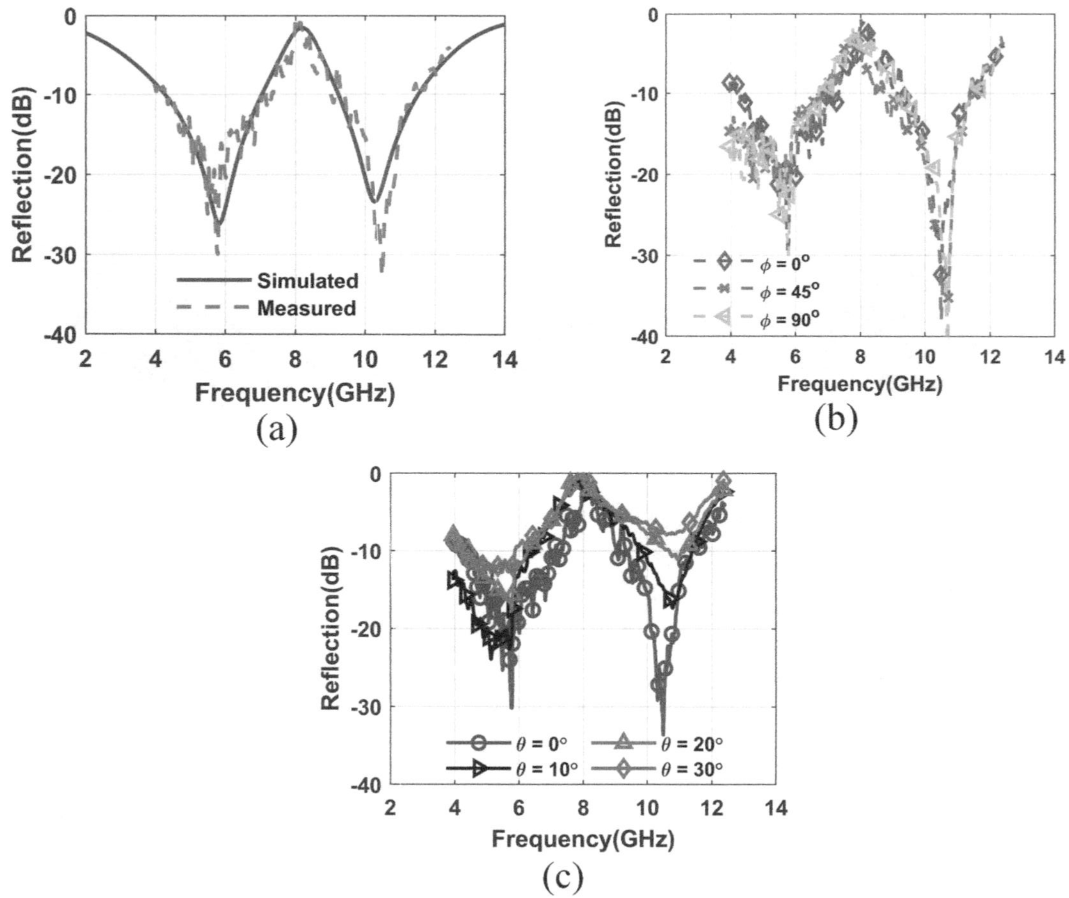

Figure 8.7 Measured reflection coefficients for the proposed AFSR structure (a) in comparison with the simulated result, (b) under different polarizations, and (c) oblique incidences [1].

with 0.8 mm thickness (dielectric constant ϵ_r = 4.4 and loss tangent tanδ = 0.02). The complete fabricated sample, measuring 315 mm × 315 mm, is shown in Figure 8.6. Surface-mounted chip resistors of 200 Ω (model CRC-W0603200RFKEA, VISHAY) are soldered into the gaps of the circular-cross elements, as depicted in Figure 8.6(a). On the reverse side of the top resistive layer, metallic circular ring resonators are patterned, as shown in Figure 8.6(b). The resistive layer is spaced from the ground plane using plastic spacers, as illustrated in Figure 8.6(c).

Reflection measurements on the developed prototype were performed in an anechoic environment using the free-space method. Standard gain horn antennas corresponding to the C, J, and X frequency bands were employed, connected to a Keysight PNA N5224B vector network analyzer. Figure 8.7(a) presents the measured reflection coefficient alongside the simulated result, where the close agreement between both confirms the experimental validity of the proposed AFSR structure. Additionally, the structure's insensitivity to polarization was verified by testing at different polarization angles, as illustrated in Figure 8.7(b). To assess angular stability, measurements under oblique incidence were also conducted (Figure 8.7(c)). The results show that stable absorption performance is maintained up to an incidence angle of 30°, after which performance begins to deteriorate across both absorption bands.

8.2 LOW-RCS CROSSED DIPOLE ANTENNA

This section introduces a dual-polarized antenna design that incorporates a polarization-insensitive AFSR structure to achieve reduced radar cross-section (RCS). The AFSR configuration is co-optimized with a dual-polarized crossed dipole antenna, which is tuned to operate near 8.2 GHz – matching the reflection notch of the AFSR. The feeding mechanism for the dipole uses a microstrip-to-broadside coupled stripline transition, as described in [2]. For reference, a conventional configuration using a crossed dipole placed above a metallic ground plane is illustrated in Figure 8.8(a), where the reflector dimensions match a 6 × 6 array of AFSR unit cells (90 mm × 90 mm). In the proposed low-RCS antenna, the same 6 × 6 AFSR array replaces the metallic reflector. A central square area on the resistive layer, sized according to the feed substrate, is removed to accommodate the dipole structure. The dipole, mounted on its substrate as shown in Figure 8.8(b), is embedded into this gap and electrically connected to the AFSR ground plane. Thus the traditional reflector is effectively substituted with the engineered AFSR surface. Figure 8.9 presents images of the fabricated prototypes for both the conventional and AFSR-integrated crossed dipole antenna configurations.

Figure 8.10 presents both simulated and measured results for the reflection coefficient (S_{11}) of the crossed dipole antenna integrated with a metallic reflector and with the AFSR surface. In both configurations, the antenna maintains a return loss better than –10 dB at the resonance frequency of

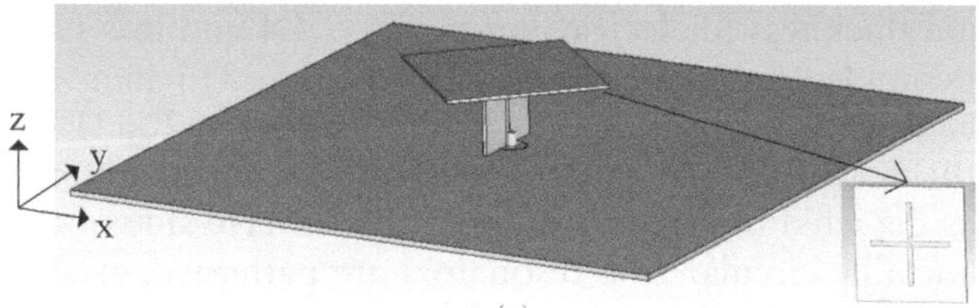

(a)

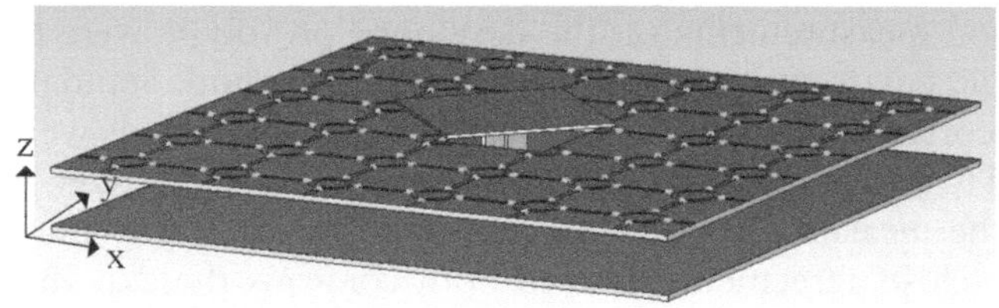

(b)

Figure 8.8 Schematic representation of the crossed dipole antenna with (a) reflector and (b) AFSR structure [1].

8.2 GHz. At this frequency, the AFSR-backed antenna achieves a gain of 5.52 dBi, which is marginally lower than the 6.35 dBi gain recorded for the metallic reflector-backed design. The radiation patterns obtained through simulation and measurement for both setups are displayed in Figure 8.11 across the *xz*- and *yz*-planes. It can be noted that the AFSR-integrated antenna exhibits radiation characteristics that closely resemble those of the conventional reflector-backed counterpart.

8.2.1 Monostatic RCS Measurement

Radar cross-section (RCS) measurements are conducted on the fabricated prototypes of the crossed dipole antenna backed by both a metallic reflector and the AFSR surface. A monostatic RCS setup is employed, utilizing the Keysight PNA N5224B, which is equipped with time-domain gating functionality. A single standard gain horn antenna is used in the same position for transmitting and receiving signals. The RCS is evaluated using the general formulation provided in Equation (8.1).

$$RCS = \frac{P_r}{P_t} \cdot \frac{(4\pi)^3 R^4}{G_t G_r \lambda^2} = K \cdot \frac{P_r}{P_t} \tag{8.1}$$

In Equation (8.1), P_r and P_t correspond to the received and transmitted power, respectively. The antenna gains for reception and transmission are denoted by G_r and G_t, while R represents the distance to the target, and λ stands for the wavelength.

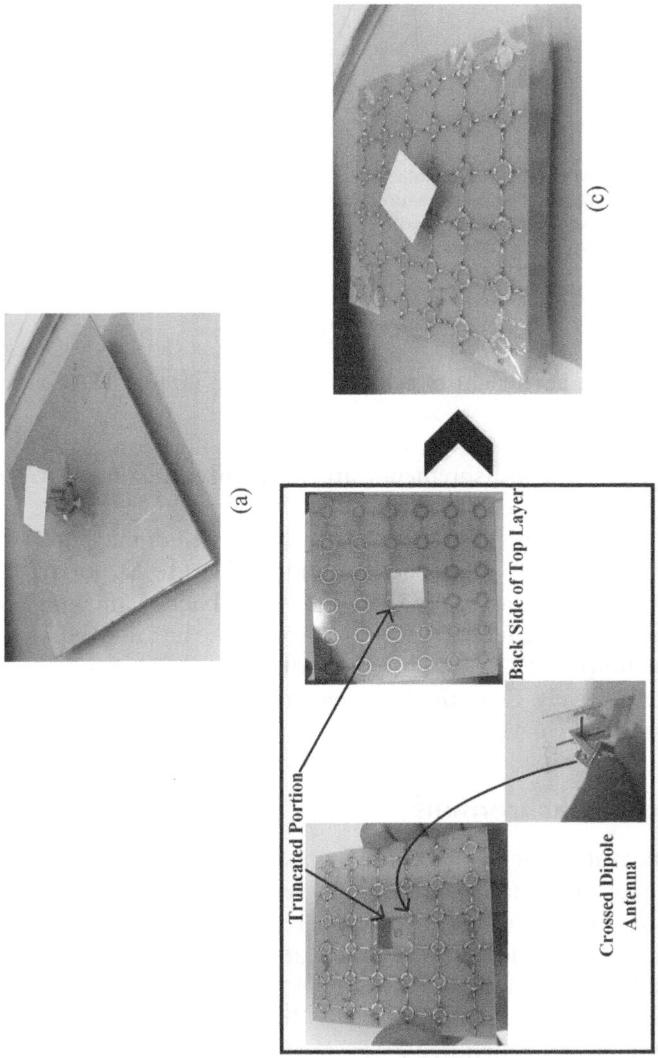

Figure 8.9 Photographs of the fabricated prototype of the crossed dipole antenna with (a) reflector, (b) truncated 6 × 6 structure, and (c) final AFSR structure [1].

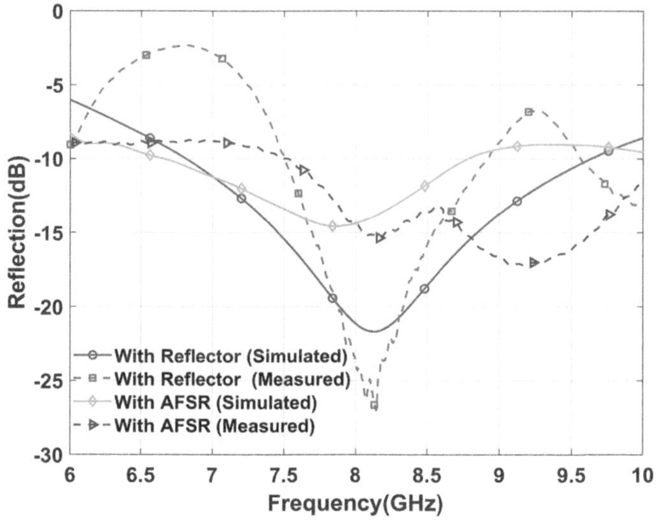

Figure 8.10 Reflection coefficient of the crossed dipole antenna [1].

The monostatic RCS measurements are normalized against a reference square PEC plate having identical dimensions to the grounded surface area of the antenna. The comparative RCS results for the crossed dipole antenna with both metallic reflector and AFSR backing under TE and TM polarized incident waves are presented in Figure 8.12. The AFSR-backed antenna demonstrates a significant RCS reduction in comparison to the metallic reflector case, with peak reductions of 27.52 dB and 38.01 dB at 10.6 GHz for TE and TM polarization, respectively. Over the frequency range 4.2–11.5 GHz, the average RCS suppression achieved is 12.51 dB for TE and 12.62 dB for TM waves.

8.2.2 Bistatic Measurement

Bistatic RCS measurement is performed by positioning the transmitting and receiving antennas at different spatial locations. In this setup, similar to the configuration illustrated in Figure 6.20 of Chapter 6, the transmitting horn antenna is oriented perpendicularly to the test structure, while the receiving antenna is rotated along a circular arc from 0° to 90°, in 10° increments, either in the clockwise or counterclockwise direction. The scattered electric field at each angular step, denoted as $E_{sca}(f,\phi)$, is mathematically expressed in Equation (8.2) [3].

$$E_{sca}(f,\phi) = S_{21,O}(f,\phi) - S_{21,FS}(f,\phi) \tag{8.2}$$

In this context, $S_{21,O}$ and $S_{21,FS}$ represent the transmission coefficients measured between the horn antennas in the presence of the object under test

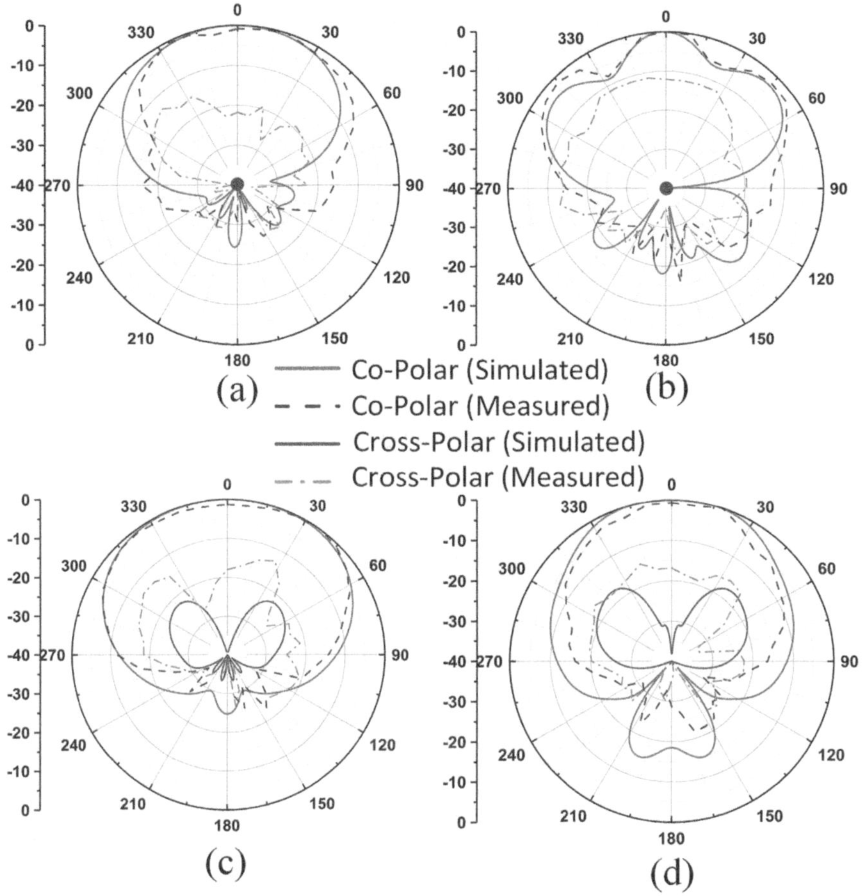

Figure 8.11 Radiation pattern of the crossed dipole antenna at 8.2 GHz: (a) **xz**-plane (with reflector); (b) **xz**-plane (with AFSR); (c) **yz**=plane (with reflector); (d) **yz**=plane (with AFSR) [1].

and in free space, respectively. The normalized total radar cross-section, denoted as $RCS_{t,\text{norm}}$, for the antenna integrated with the AFSR structure is determined by integrating the power of the scattered fields, as expressed in Equation (8.3).

$$RCS_{t,\text{norm}} = \frac{\int \left| E_{\text{sca},AFSRA}(f,\phi) \right|^2 d\phi}{\int \left| E_{\text{sca},RA}(f,\phi) \right|^2 d\phi} \tag{8.3}$$

where $E_{\text{sca},AFSRA}$ and $E_{\text{sca},RA}$ represent the scattered electromagnetic fields corresponding to the AFSR-backed antenna and the standard reflector antenna, respectively.

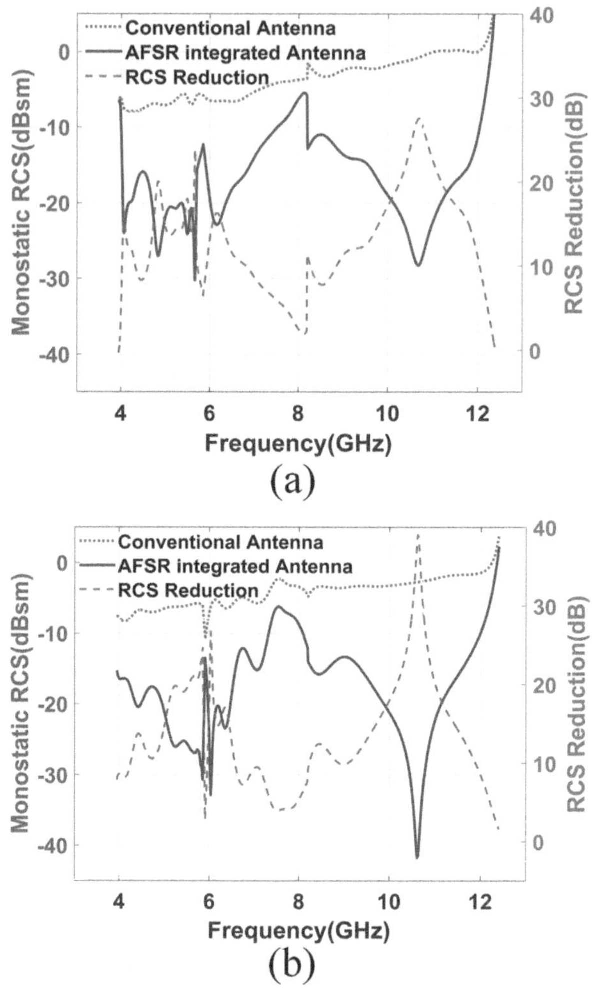

Figure 8.12 Monostatic RCS of the AFSR-integrated antenna in comparison with the conventional-reflector-based antenna for (a) TE and (b) TM incident wave [1].

Figure 8.13 presents the measured normalized total RCS ($RCS_{t,\mathrm{norm}}$) for the crossed dipole antenna integrated with the AFSR structure, compared to the traditional metal reflector-backed design. It is evident that the antenna utilizing the AFSR configuration achieves a notable decrease in total RCS relative to the metallic reflector counterpart. Across the frequency range 4.2–11.5 GHz, the average normalized RCS values are approximately 0.197 and 0.198 for TE- and TM-polarized waves, respectively, indicating an overall RCS reduction of nearly 80.24% for both polarization types compared to the standard reflector-based antenna.

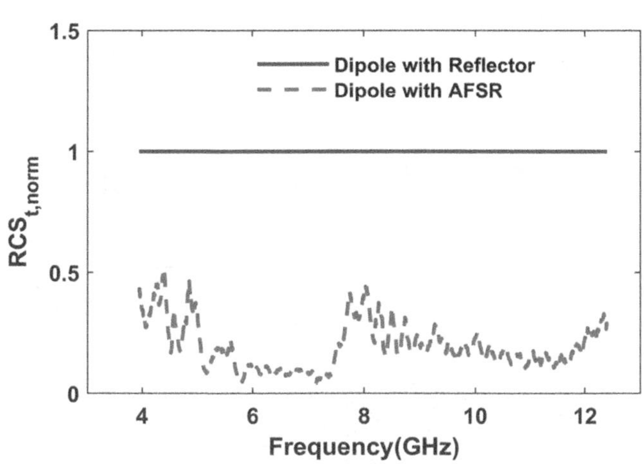

(a)

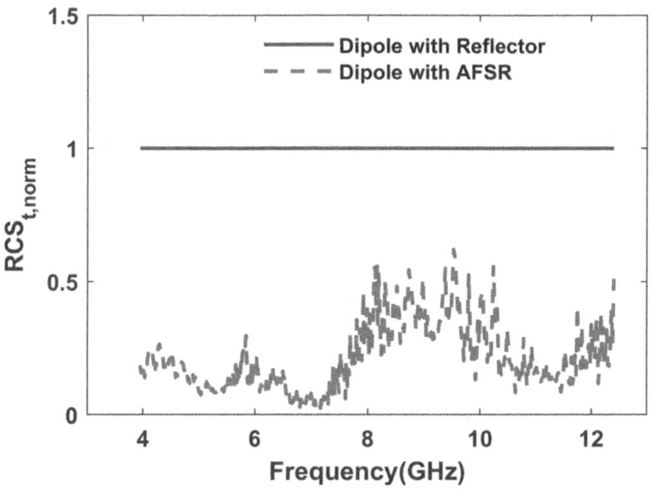

(b)

Figure 8.13 Normalized total bistatic RCS of the AFSR-integrated antenna with reference to the conventional reflector-based antenna for (a) TE and (b) TM incident wave [1].

Table 8.1 compares the proposed AFSR-based crossed dipole antenna with other existing low-RCS antenna designs from the literature. Unlike previously reported works [4–10], which primarily focus on single-polarized antennas, this work demonstrates RCS reduction for a dual-polarized crossed dipole for the first time. The antenna structure presented here is more compact than those discussed in [4, 6, 7], and it features reduced structural

Table 8.1 Comparative Performance of the Proposed AFSR-Integrated Antenna with the Recently Reported Low-RCS Antennas [1]

Ref.	*Antenna Structure*	*Operating Frequency*	*Size*	*Lumped Elements in the Structure*	*10 dB Monostatic RCS Bandwidth*	*Average Monostatic RCS Reduction*	*Average Total Bistatic RCS Reduction*
[4]	Dipole + AFSR	8.7 GHz	$3.65\lambda \times 3.65\lambda$	200	26% (lower) and 16.7% (upper)	NA	NA
[5]	Monopole/dipole + 3D AFSR	5.5 GHz	$2.56\lambda \times 1.54\lambda$	196	64.7% (lower) and 42.4% (upper)	NA	NA
[6]	Patch array antenna + AFST	10 GHz	$7.2\lambda \times 7.2\lambda$	96	60.3% (lower) and 20.4% (upper)	NA	NA
[7]	Patch antenna + Absorptive/ diffusive FSR	10 GHz	$10.66\lambda \times 10.66\lambda$	512	90% (lower) and 27.6% (upper)	NA	NA
[8]	Patch antenna + ASFT	7.5 GHz	$1.5\lambda \times 1.5\lambda$	128	6.9% (lower) and 39.3% (upper)	NA	NA
[9]	Slot array antenna + 3D AFST	6.15 GHz	$1.96\lambda \times 1.96\lambda$	192	87.2% (lower) and 45.8% (upper)	NA	NA
[10]	Cross-shaped vivaldi antenna	1.8–6 GHz	$0.58\lambda \times 0.58\lambda$	NA	NA	4.3 dB (TE) and 5 dB (TM)	NA
This work	Crossed dipole + AFSR	8.2 GHz	$2.46\lambda \times 2.46\lambda$	144	38.59% (lower TE)/42.9% (lower TM) and 30.75% (upper TE), 27% (upper TM)	12.51 dB (TE) and 12.62 dB (TM)	80% (TE and TM)

Note: λ is wavelength at the operating frequency of antenna.

complexity by employing fewer lumped components compared to [4, 5, 7, 9]. Additionally, the proposed antenna offers a broader −10 dB RCS reduction range in the lower frequency band than designs in [4, 8] and a wider upper band than those in [4, 6, 7]. The average RCS suppression achieved in this design is also notably superior to that reported in [10]. Importantly, the RCS mitigation is realized using a planar 2D AFSR configuration, unlike the 3D frequency selective structures employed in [5, 9]. Moreover, the scattered field analysis in this work includes angular measurements, leading to an experimentally validated average bistatic total RCS reduction of approximately 80% for both TE and TM polarized waves, a capability not demonstrated in earlier studies [4–10].

8.3 CONCLUSION

This chapter introduces a method for reducing radar cross-section (RCS) by integrating an antenna with an AFSR-based surface. Initially, an AFSR structure is developed that features a reflective notch at 8.2 GHz, positioned between two broad absorption ranges spanning from 4.2–7.0 GHz and 9.2–11.5 GHz. This is accomplished by etching a ring-shaped bandstop resonator on the back side of the upper resistive substrate. A crossed dipole antenna, tuned to operate at the reflection notch frequency, is embedded in a centrally truncated region of a 6 × 6 array of the AFSR unit cells. When compared to a traditional metallic reflector configuration, the integrated design achieves average monostatic RCS reductions of 12.51 dB for TE polarization and 12.62 dB for TM polarization over the 4.2–11.5 GHz band. Additionally, the experimental results confirm an approximate 80% reduction in total bistatic RCS relative to the conventional reflector-backed antenna.

REFERENCES

1. M. M. Zargar et al., "Low RCS Dual-Polarized Crossed Dipole Antenna Co-Designed With Absorptive Frequency-Selective Reflection Structure," IEEE Access, vol. 10, pp. 118806–118814, 2022.
2. K. Saurav, D. Sarkar and K. V. Srivastava, "CRLH Unit-Cell Loaded Multiband Printed Dipole Antenna," IEEE Antenn. Wirel. Propag. Lett., vol. 13, pp. 852–855, Apr. 2014.
3. P. Alitalo, A. E. Culhaoglu, A. V. Osipov et al., "Experimental Characterization of a Broadband Transmission-Line Cloak in Free Space," IEEE Trans. Antenn. Propag., vol. 60, no. 10, pp. 4963–4968, Oct. 2012.
4. P. Mei, X. Q. Lin, J. W. Yu, A. Boukarkar, P. C. Zhang and Z. Q. Yang, "Development of a Low Radar Cross Section Antenna With Band-Notched Absorber," IEEE Trans. Antenn. Propag., vol. 66, no. 2, pp. 582–589, Feb. 2018.

5. H. Huang, Z. Shen and A. A. Omar, "3-D Absorptive Frequency Selective Reflector for Antenna Radar Cross Section Reduction," IEEE Trans. Antenn. Propag., vol. 65, no. 11, pp. 5908–5917, Nov. 2017.
6. Q. Chen, M. Guo, D. Sang, Z. Sun and Y. Fu, "RCS Reduction of Patch Array Antenna Using Anisotropic Resistive Metasurface," IEEE Antenn. Wirel. Propag. Lett., vol. 18, no. 6, pp. 1223–1227, Jun. 2019.
7. B. Zhang, L. Li, C. Jin, Q. Lv and R. Mittra, "Wideband Low RCS Antenna Based on Hybrid Absorptive-Diffusive Frequency Selective Reflector," IEEE Access, vol. 9, pp. 77863–77872, Jun. 2021.
8. W. Yu, Y. Yu, W. Wang, X. H. Zhang and G. Q. Luo, "Low-RCS and GainEnhanced Antenna Using Absorptive/Transmissive Frequency Selective Structure," IEEE Trans. Antenn. Propag., vol. 69, no. 11, pp. 7912–7917, Nov. 2021.
9. Y. Yu, W. Yu, G. Xie, J. Tong and G. Q. Luo, "Wideband RCS Reduction of Slot Array Antenna Utilizing 3-D Absorptive Frequency-Selective Structure," IEEE Antenn. Wirel. Propag. Lett., vol. 21, no. 1, pp. 64–68, Jan. 2022.
10. K. Zhang, R. Tan, Z. H. Jiang, Y. Huang, L. Tang and W. Hong, "A Compact, Ultra-Wideband Dual-Polarized Vivaldi Antenna with Radar Cross Section Reduction," IEEE Antenn. Wirel. Propag. Lett., vol. 21, no. 7, pp. 1323–1327, Jul. 2022.

Index